Institutions, Technology and Development in Africa

An extensive literature has demonstrated that technologies in sub-Saharan Africa are largely inappropriate: that is, that they are typically capital- and import-intensive rather than labour- and local input-intensive. These technologies have created a pattern of development that is highly unequal, with widespread unemployment and under-employment. In this literature, however, relatively little attention has been paid to the institutions that govern the generation, adoption and use of technology.

This book draws on historical analysis and case studies to evaluate how institutions in different countries, including those in Africa itself, generate technologies that vary in their characteristics and suitability for the region. Through these case studies, insight is gained into the characteristics of 'appropriate' institutions that might underlie a more balanced pattern of technology and development than currently exists. The findings of the book clearly confirm a major tenet of institutionalist theory: namely, that institutions developed in one set of circumstances are unlikely to be appropriate to conditions in a markedly different set.

This book will be of interest to economists, social historians and anyone with an interest in modern African development.

Jeffrey James has researched and taught in South Africa, England, the USA and the Netherlands. He has written extensively in the area of technology and development, including issues related to emerging digital technologies. Much of his work in this area relates to sub-Saharan Africa.

Routledge Studies in Development Economics

For more information about this series, please visit www.routledge.com/series/
SE0266

Institutions, Technology and Development in Africa

Jeffrey James

Routledge
Taylor & Francis Group

LONDON AND NEW YORK

First published 2019
by Routledge

2 Park Square, Milton Park, Abingdon, Oxfordshire OX14 4RN
52 Vanderbilt Avenue, New York, NY 10017

Routledge is an imprint of the Taylor & Francis Group, an informa business

First issued in paperback 2020

British Library Cataloguing-in-Publication Data
A catalogue record for this book is available from the British Library

Library of Congress Cataloging-in-Publication Data
A catalog record has been requested for this book

ISBN: 978-1-138-54582-3 (hbk)
ISBN: 978-0-367-66378-0 (pbk)

Typeset in Bembo
by Apex CoVantage, LLC

Contents

Figures

Tables

Acknowledgements

I am especially grateful to Natalie Tomlinson at Routledge for her support and encouragement at a time when it was most needed. She is a great asset to her firm and has been since I first began publishing with Routledge many years ago. I also received able support from Lisa Lavelle in her capacity as editorial assistant.

For the typing and editing of the manuscript sincere thanks go to Ailsa Rainer, René Wijnen and Lisette Barten of Tilburg University. René's aptitude for organisation helped to offset my severe deficiencies in this area. At the Copyright Clearance Center, Mona Are provided invaluable assistance. I regret to say that I was refused permission, by Sacmeq, to use certain data on Francophone African countries without any explanation under their author re-use policy.

Chapter 5 is from Jeffrey James (2014), 'Internet use, welfare and well-being: evidence from Africa', *Social Science Computer Review*, 32, 6:715–727, by permission of Sage.

Chapter 6 is from Jeffrey James (2012), 'Institutional and societal innovations in IT for developing countries', *Information Development*, 28.3:183–188, by permission of Sage.

Chapter 7 is from Jeffrey James (2015), 'Macro consequences of the One Laptop per Child project', *Journal of International Development*, 27,1:144–146, with permission of Wiley.

Chapter 8 is from Jeffrey James (2008), 'Sharing mechanisms for IT in developing countries, social capital and quality of life', *Social Indicators Research*, 94,1:43–59, by permission of Springer.

Chapter 9 is from Jeffrey James (2018), 'A sequential analysis of the welfare effects of mobile phones in Africa', first published online February 4, *Social Science Computer Review*, by permission of Sage, under their author re-use policy.

Table 1.1 is from Jeffrey James (2006), 'Bridging the source of the digital divide', paper presented at the Digital Opportunity Forum, held in Seoul, Korea from August 31 to September 1, 2006.

Table 2.1 is made up of excerpts from A. Hartmann and J. Linn (2008), 'Scaling up: a framework and lessons for development effectiveness from literature and practice', Brookings Institution, with permission.

Table 3.1 and 3.2 are from H. Pack (1982), 'Aggregate implications of factor substitution in industrial processes', *Journal of Development Economics*, 11, 1:1–37, by permission of Elsevier.

Table 3.3 is from F. Tregenna (2012), 'Sectoral labour-intensity in South Africa', Nedlac. Labour-Intensity Report, by permission of the author.

Table 3.4 is from Research ICT Africa (2011), survey data, with permission.

Table 3.5 is from UN (2015), *World Population Prospects: The 2015 Revision*, ©2015. Reprinted with the permission of the United Nations.

Table 3.5 is from Edwards, L. and Jenkins, R. (2015), 'The impact of Chinese import penetration on the South African manufacturing sector', *The Journal of Development Studies*, 51,4, reprinted under Creative Commons Attribution License, http://creativecommons.org/licenses/by/4.0.

Tables 4.1 to 4.5 are from World Bank Indicators, 2017.

Table 4.6 is from K. Watkins (2013), 'Too little access, not enough learning: Africa's twin deficit in education', Op-ed, Brookings Institution, Jan. 16, with permission of Brookings and the PASEC institution.

Table 4.7 is from World Bank Indicators, 2017.

Table 4.8 is from World Bank Indicators, 2015.

Figure 5.1 is from Jeffrey James (2014), 'Internet use, welfare and well-being: evidence from Africa', *Social Science Computer Review*, 32, 6, 715–727, with permission of Sage.

Tables 5.1, 5.3, 5.4, 5.5, 5.7, 5.8, 5.9 and 5.10 are from a survey conducted by Research ICT Africa 2011 with permission.

Table 5.6 is from PEW Internet (2012) and Research ICT Africa (2011), with permission. PEW bears no responsibility for the conclusions reached on the basis of the data.

Figure 6.1 is from James, J. and Stewart, F. (1981). 'New products: a discussion of the welfare effects of the introduction of new products in developing countries', *Oxford Economic Papers*, 33, 1:81–107. By permission of Oxford University Press.

Table 6.1 is from Gomez, R., Ambikar, R. and Coward, C. (2008). Libraries, Telecenters and Cybercafés: A Study of Public Access Venues Around the World, a conference paper, IFLA, https://archive.ifla.org/IV/ifla74/papers/107-Coward_Gomez_Ambika-en.pdf. With permission of IFLA (The International Federation of Library Associations and Institutions).

Table 7.1 is from World Bank, Nationmaster, UNESCO data and own calculations.

Table 8.1 is from Jeffrey James (2007), 'From origins to implications: key aspects in the debate over the digital divide', *Journal of Information Technology*, 22, 3:284–295, with permission of Springer.

Table 8.2 is from World Bank, ICT-at-a-glance tables.

Table 9.2 is from World Bank Indicators, 2015.

Table 9.3 is from Jeffrey James (2016), *The Impact of Mobile Phones on Poverty and Inequality in Developing Countries*, Springer, with permission.

Table 9.4 and 9.5 are from Bayes, A., von Braun, T. and Akhter, R. (1999). 'Village payphones and poverty reduction: insights from a Grameen Bank initiative in Bangladesh', discussion papers on development policy, number 8, Centre for Development Research, Bonn, published with permission.

1 Introduction

Theories of institutions

This book introduces and applies the idea of *appropriate institutions*: that is, institutions which enable the benefits of technology to be more widely spread among the less-advantaged groups in African countries.[1] Though there is by now a vast literature on appropriate technology, the analysis of institutions has lagged far behind. That is, while the debate about appropriateness has concerned itself with the desirability of particular technologies, my concern here is more with the appropriateness of the institutions that are concerned to generate, adopt and scale up various technologies. As I see it, the debate up till now has been too heavily focused on the one side – the technologies – and not enough on the other – the institutions. Yet, as Acemoglu and Robinson (2010) have persuasively argued, it is the latter that tend to dominate the reasons for economic growth in a variety of countries. And in the development context, Shirley has argued, symmetrically, that 'the largest barriers to development arise from a society's institutions – its norms and rules' (Shirley, 2010: 1). My approach to this core topic of the book is embedded in certain aspects of institutional economics, as described in the following section. Thereafter, I deal with a summary of the technologies and institutions that comprise the remaining chapters.

Theories of institutions

According to Hodgson (2006: 2), institutions may be defined as 'systems of established and prevalent social rules that structure social interactions. Language, money, law, systems of weights and measures, table manners, and firms (and other organizations) are thus all institutions'.

Hodgson is also at pains to distinguish between so-called institutionalist theory and neoclassical economics, as is clear from the following quotation. First,

> There is a degree of emphasis on institutional and cultural factors that is not found in mainstream economic theory. Second, the analysis is openly interdisciplinary, in recognizing insights from politics, sociology, psychology, and other sciences. Third, there is no recourse to the model of the rational, utility-maximizing agent. Insomuch as a conception of the individual agent is involved, it is one which emphasizes both the prevalence of habit and the

possibility of capricious novelty. Fourth, mathematical and statistical techniques are recognized as the servants of, rather than the essence of economic
theory. Fifth, the analysis does not start by building mathematical models:
it starts from stylized facts and theoretical conjectures concerning causal
mechanisms. Sixth, extensive use is made of historical and comparative
empirical material concerning socio-economic institutions.

(Hodgson, 2006: 174)

To this list should perhaps be added a major tenet of what is known as the 'old'
institutional economics, associated with the likes of Veblen, Commons and, more
recently, Myrdal – that is, the notion that institutions from one environment are
unlikely to fit in easily with conditions in a totally different environment, as has
already been amply demonstrated in the transfer of technologies from developed
to developing countries.[2] With respect to the new information technologies
that are the subject of the second part of this book, for example, Gough and
Grezo observe that:

> the ways in which mobiles are used, valued and owned in the develop-
> ing world are very different from the developed countries. More attention
> should be paid to the characteristics of how people actually do use phones in
> the developing world in policy debates on increasing access to Information
> and Communication Technology (ICT). It is wrong to simply extrapolate
> our developed world models of needs and usage patterns to poorer nations.
>
> (Gough and Grezo, 2005: 1)

Institutionalist theory as a framework

The chapters below conform in many ways to these characteristics of insti-
tutionalism. For one thing, they focus explicitly not just on technology and
institutions, but also on the influence of the latter on the benefits conveyed by
the former, to particular income groups in African countries. Another reason
is the relatively heavy use that is made of disciplines other than economics. I
am referring here, for example, to politics, sociology and anthropology. Often,
'extensive use is made of historical and comparative empirical material concerning socio-economic institutions' (Hodgson, 1998: 173).

Then again, at no point does my analysis follow the neoclassical tradition of
taking the individual as given, or as an agent of maximisation. Finally, perhaps
the main theme of the chapters that follow is the difference in institutions
between rich and poor countries, especially as those institutions bear on the
well-being of individuals with relatively low incomes in the latter.

Yet another respect in which the analysis below departs from neoclassical
economics is that the benefits of commodities and technologies are not taken
to accrue to the user at the point of purchase. Rather, such gains are thought to
be dependent on a wide range of factors that emerge *after* the point of purchase.
Much depends, for example, on how the good is actually used.

In the context of medicinal drugs, this might mean whether they are used as directed, or whether, for example, they are misused. Or, in the area of food, welfare depends on more than just nutrition (Sen, 1985). Rather, it depends, among other things, on whether other persons are also present, as opposed to a situation where eating is a solitary process (Sen, 1985). More generally, welfare depends heavily on the context, a theme that re-occurs throughout the book.

That a local institution in a developing country can have a major impact on society is perhaps best illustrated by the Grameen Telecom villages phone project (VPP) in Bangladesh. In each village where it operates, a 'phone lady' is appointed and receives a loan from the project to buy a mobile phone and rent it out to villagers. The VPP has undeniably been a great success 'by bringing IT to the poor and giving telecom access to rural people for whom telephones were a luxury rather than a basic right. By 2007, the program spread to more than 50,000 villages' (Yusuf and Alam, 2011: 36). Moreover, it has been estimated that some 30,000 jobs have been created for telephone ladies, who might otherwise have gone unemployed.

Also well known in this regard is the Community Phone Shop in South Africa, which was specifically designed to operate as a local phone booth, containing five to ten phones and located in disused containers and informal 'spaza' shops. These institutions operate widely as public mobile phone services in townships and other disadvantaged areas (see below).

National innovation systems

Another aspect of institutionalist theory that bears mention here is the idea of a so-called national innovation system. It has been defined by Freeman as 'the network of institutions in the public and private sectors whose activities and interactions initiate, import, modify and diffuse new technologies' (Freeman, 1987: 1). The idea is especially relevant to the case studies in Chapter 2, which address themselves to the interactions between participating actors, whether they be small or large-scale, public or private, national or international. In most cases, such institutional interactions in Africa tend not to produce the outcomes that were hoped for from them, but in these four case studies, the result of scaling up was largely successful. Using some of the literature on the topic, I pay particular attention to the striking degree of success that was achieved in the scaling up of labour-intensive methods of rural road construction in Kenya, where the outcome of institutional interactions turned out to be highly favourable (see Chapter 2). An important discussion has subsequently addressed itself to the replicability of this and other examples, in different parts of the region.

Chapters 3 and 4

These two chapters remain squarely within the remit of institutionalist theory. Chapter 3 describes how changing international institutions bear on the appropriateness of the technologies that are used in Africa. On the one hand, I show how institutional resources for research on appropriate technology there have declined over the past 20 years or so. On the other hand, there are countervailing forces in the form of a rapidly growing number of Chinese and Indian firms. My hypothesis is that these firms tend to promote more appropriate technologies than either institutions from developed countries, or local firms in comparable industries.

The reasoning here is that techniques in these two large developing countries are generated against a background of socio-economic and institutional factors that are more akin to African than developed-country conditions. And there is already some evidence (as noted in a later chapter) that private Chinese firms tend to be associated with relatively appropriate technologies. Much research, however, remains to be done on this important topic.

Chapter 4 is about capabilities rather than technologies, but it is based nevertheless on the institutionalist notion that institutions from developed countries are often unsuitable for conditions prevailing in much poorer countries, such as those in Africa. In this case, the institutions in question are the components of technological indexes that are designed very much with developed countries in mind. Thus, they include measures such as R&D, patents and high-tech exports, which have very little to do with production structures in most African countries, and as a result are unable to discriminate adequately between them.[3] What I propose instead is an index designed specifically for African countries, that is based on more fundamental elements such as literacy, numeracy and vocational training. According to this revised index, several African countries perform much better than the rest of what is admittedly a small sample.

Part II

The chapters in the second part of the book are linked together by the fact that they are all concerned with information technology, and with means of access to this technology, that vary from one group of countries to another. In developed countries, for example, the overwhelming mode of access to the Internet and mobile phones is through ownership. In developing countries, by contrast (particularly with regard to rural areas), ownership is frequently not a viable option, because incomes for many inhabitants are much too low to afford this form of access. Many of the chapters in this part of the book, therefore, are concerned with institutions that enable the benefits of the new technology to be brought by other means to excluded groups. They are also concerned, in other words, with what I referred to above as 'appropriate institutions'.

Appropriate institutions and information technology

In order to reach the groups which cannot rely on ownership as a means of access to information technology, there are, broadly, two alternative institutional forms available. One allows the use of IT without ownership by the users, while the other permits the benefits of technology to be reaped even without any individual use of it whatsoever:

For reasons that have to do with differences in the user capabilities they require, the first form of change applies best to mobile phones, whereas the second is most obviously suited to extracting the benefits of the Internet in rural areas of developing countries.

(James, 2006: 6)

To be specific, while mobile phones make virtually no demands on user skills (not even literacy), the Internet requires a host of skills, ranging from computer literacy and linguistic skills to knowledge of the more demanding aspects of computing. Such differences as these are important because they indicate the direction of institutional change that is required in relation to mobile phones, on the one hand, and the Internet, on the other. For whereas with respect to the former type of IT, the need is to extend the limits imposed by ownership using sharing mechanisms of one kind or another, the Internet requires innovations that bring the benefits to the rural majority, without any need for individual use of the technology (given the vast gap between the user capabilities available to this group and those that are actually required) (James, 2006: 6).

Table 1.1 contains a taxonomy of the previous discussion with commercial and non-commercial modes of institutional change in the case of mobile phones, and face-to-face vs distance intermediation, in the case of the Internet.

Chapters 5 to 9 provide further insight into the relationships that have just been portrayed. Chapter 9 is perhaps especially important in this regard because

Table 1.1 Illustrative cases of institutional change in mobile phones and the Internet

Institutional change to expand users	*Institutional change to derive benefits without use*
Mobile phones	*The Internet*
(a) *Non-commercial*	(a) *Face-to-face intermediation*
• Sharing a mobile phone by the friends and family of owners.	• Rural Internet kiosks that are operated by people familiar with the technology and the local community (enabling poor, illiterate rural inhabitants to have e-mails sent and government documents received).
(b) *Commercial*	
• Buying time from vendors situated in villages, small towns, roadside kiosks.	(b) *Distance intermediation*
• People who cannot afford a mobile phone use prepaid cards to make calls from a handset belonging to someone else.	• Community radio stations that transmit, translate and contextualise information from the Internet for the benefit of listeners (even those living in remote, rural areas). For example, in one project, there are experts such as doctors, teachers and lawyers in the radio station who, on air, use the Internet to provide information in their own area of expertise to the listening public. The information is translated, contextualised, and otherwise made accessible to the audience.
• Renting prepaid cards (users provide the code belonging to a card in order to make a call at a payphone and pay the vendor for just the number of units that were used. The vendor, in turn provides the code to another client and the process continues until the card is depleted).	

it shows that leapfrogging (or bypassing an older technology) amplifies the gains from sharing mobile phones in African countries, i.e., from appropriate institutions. For while the phones confer a crucial means of communication on those who suffer a pronounced disadvantage in this respect, the gains are rendered greater from this same lack of alternatives.

Imagine, for the sake of argument, a remote farmer who needs to communicate with a supplier from another region. Possessed of fixed-phones or other means of communication, such as adequate public transport, the gains to the farmer from mobile phones will be more muted than in the case where no alternatives are available. This mechanism has been confirmed in case studies and also lies behind the finding that the growth effects of mobiles tend to be greater in poor rather than rich countries.

Note also that the welfare effects of mobile phones cannot be assessed only with regard to outgoing calls, for much also depends on the possibility of getting incoming calls in a situation where the phones are shared. There are two considerations here. One is whether the party who receives the incoming call is notified about it, as occurs, for example, in the Grameen Telecom project in Bangladesh. Or, in the more common situation where sharing takes place among close friends and family, it is sometimes the case that incoming calls are passed on to the recipient, though this depends on the extent of goodwill and trust that exists within the group. The second consideration has to do with the salience of the missed call itself. This can range from the trivial, to an important message about, say, the health of a relative. The size of the negative welfare effects will vary accordingly (and the overall effect will be less than the situation where both outgoing and incoming calls are possible).

Chapter 7 is a case study of the well known One Laptop per Child (OLPC) project in primary schools in three developing countries, including one from sub-Saharan Africa. The study is unusual in that it deals with the appropriateness of both the technology (the laptops) and the institutions that govern their use. The so-called XO computer is judged to be appropriate, not only because it is designed more generally for the conditions prevailing in such countries (for one thing, because it is more rugged than the typical developed-country product). As far as the associated institutions are concerned, however, I find that the practice of giving a laptop to each and every student is often inappropriate, because when widely used, it tends to cause an imbalance in the national education budget, an imbalance which crowds out other desirable activities in the sector. On the basis of a simple formula, this brief chapter shows that the degree of imbalance thus created will tend to be most severe in the poorest developing countries, many of which are located in Africa. In such countries, especially, it seems more appropriate if some degree of institutional sharing among students takes place. Here, therefore, as in some of the earlier chapters, sharing appears to be an appropriate means of bringing the benefits of information technology to those with low incomes in rural areas.

Chapter 8 is distinctive, among other reasons, because it provides a detailed comparison of sharing mechanisms across all the main types of information technology: namely, the Internet, mobile phones and computers. Numerous examples are provided for each item in the taxonomy, so as to afford policy-makers with an informed set of options in different circumstances. It bears emphasising that most of the options are local in nature, and differ, as might be expected, from those found in the developed countries (which are generally not designed with poor-country circumstances in mind). This chapter is also distinctive in that it draws on the concept of social capital, which, according to some authors, forms part of institutionalist theory. The point is that the value of this concept can be increased by information technology in developing countries (where trust and trustworthiness are generally lower than in developed countries).[4] 'In fact, substantial evidence demonstrates that social norms prescribing cooperative or trustworthy behaviour have a significant impact on whether societies overcome obstacles to contracting and collective action that would otherwise hinder their development' (Keefer and Knack, 2003: 1).

Chapter 9 is a synthesis in that is seeks to integrate areas of the literature that are usually kept apart from one another. It does so by means of a sequential analysis in which effects at one stage influence those at later stages. For example, mobiles generated in China and India may have a quite different welfare effect on poor users in Africa, because they tend to be much cheaper than alternatives from most other countries (though there may often also be some offsetting quality differences). Such a possibility, however, has thus far attracted much less research attention than it deserves. Chapter 9 also differs from other chapters in that it focuses on SIM cards and not just mobile phones. In parts of Africa, for example, these cards are being traded in the absence of mobile phone ownership. To this extent, the benefits of information technology are being extended beyond sharing of just the mobile phone.

The current chapter has introduced the central idea of this book, namely, the 'appropriate institution', which serves to enhance the benefits available from technology to low-income users in Africa and beyond. While there has been a vast literature on the topic of appropriate technology, this body of writing has tended to ignore the institutional context in which technologies are generated, adopted, adapted and used. I have suggested that an analysis of these issues be undertaken on the basis of the type of institutionalist theory associated with Veblen, Commons and Myrdal.

This means, among other things, an openness to a variety of different disciplines; a refusal to take individuals as given; a reluctance to accept rich-country institutions as being necessarily suitable for low-income areas of Africa; and a need to focus on empirical material and case studies of particular institutions. Unlike traditional neoclassical economics, moreover, it is assumed that welfare derived from technology does not occur at the point of purchase. Rather, it depends on the context: what other goods the user possesses and how the technology is actually applied (which depends partly on skills). These ideas, finally, are pursued across a range of technologies, from infrastructure and manufacture to various types of information technology, and include technological capabilities as well as technology choice.

Notes

1 Appropriate institutions bear the same close relationship to poverty and inequality as do appropriate technologies.
2 See Stewart (1977).
3 Yet precisely such discrimination between poor countries is an avowed goal of the Technology Achievements Index of the United Nations Development Programme (UNDP).
4 For the detailed estimates, see Knack and Keefer (1997).

References

Acemoglu, D. and Robinson, J. (2010). The role of institutions in growth and development, *Review of Economics and Institutions*, 1,2:2–33.

Freeman, C. (1987). *Technology Policy and Economic Performance: Lessons from Japan*, London: Pinter.

Gough, N. and Grezo, C. (2005). Introduction, in *Africa: The Impact of Mobile Phones*, The Vodafone Policy Paper Series, number 2, London.

Hodgson, G. (1998). The approach of institutional economics, *Journal of Economic Literature*, 36,2:166–192.

Hodgson, G. (2006). What are insitutions? *Journal of Economic Issues*, 40, 1:1–25.

James, J. (2006). *Bridging the source of the digital divide*. Paper Presented for the Digital Opportunity Forum, Seoul, Korea.

Keefer, P. and Knack, S. (2003). Social Capital, Social Norms and the New Institutional Economics, MPRA paper, 25025, University Library of Munich, Germany.

Knack, S. and Keefer, P. (1997). Does social capital have an economic payoff? A cross-country investigation, *The Quarterly Journal of Economics*, 112,4:1251–1288.

Sen, A. (1985). *Commodities and Capabilities*, Amsterdam: North Holland.

Shirley, M. (2010). *Institutions and Development*, Cheltenham: Elgar.

Stewart, F. (1977). *Technology and Underdevelopment*, London: Macmillan.

Yusuf, M. and Alam, Q. (2011). Empowering role of the village phone program in Bangladesh: In retrospect, in prospect, *Journal of IT Impact*, 11,1:35–50.

Part I
Non-digital technologies

2 Scaling up pilot projects in Africa

Four cases

This chapter begins with the observation that small-scale, labour-intensive technologies in Africa are only very rarely scaled up to the regional or national level.[1] But it is also true that there are some exceptions to this general pattern, i.e., cases in the region where individual technology projects are successfully scaled up to higher levels of aggregation. The purpose of the chapter, accordingly, is to explain these unusual cases, in the light of the then-prevailing institutional framework.

As a first step in this direction, I begin with a discussion of how the features of appropriate technology influence the type of institutions that are likely to help or hinder the process. For example, this type of technology is closely related to low incomes, rural development and job creation, and institutions which are concerned with influencing these goals will tend to have a positive influence on scaling up. The converse is true of institutions that neglect or negatively influence such goals (e.g., some types of institutions that are responsible for granting foreign aid).[2]

Then, in an effort to further understand the successful case studies below, I present some of the findings from the general scaling-up literature, which has grown quite rapidly in recent years.[3] Though not all these findings are relevant to all the case studies, they do nevertheless introduce some important political and sociological variables into the discussion (such as, for example, the degree of political support for, and commitment to, the scaling-up process by different institutions and individuals).[4]

The case studies can be divided broadly into two groups: namely, those that deal with road construction in two African countries, and those dealing with the scaling up in the region of an oil press and an improved charcoal stove. The first group is concerned mainly with the public sector's role in labour-intensive road construction, whereas the second focuses more on the role of the market in producing and distributing the improved products (though in this case too, of course, there are certain state and other institutions that guide the market).

Institutions for scaling up[5]

Centralised state institutions

From an empirical point of view, centralised institutions of the state do not, as a rule, favour the choice of scaling up of appropriate technology projects. They are given instead to capital- and import-intensive techniques, whose beneficiaries are typically allies of the central state such as large-scale firms, foreign machinery suppliers and Western multinationals and aid donors.

Then too there is a major problem with development finance institutions in African and other developing countries, which do not normally view themselves as technology institutions.[6] The result is that foreign finance leads, indirectly, via foreign consultants and aid donors, to large-scale, capital-intensive projects, which are often also highly inefficient (James, 1995).

Nor, in this process, is there typically any point at which considerations about technologies present themselves. For once a selection has been made at the design stage of the project cycle, there is no later point at which this decision is questioned, even at the appraisal stage (Jéquier and Hu, 1989). In theory, of course, appropriate technologies might be selected at the design stage, but this tends to occur only very infrequently, since the decision is usually made by Western consultants, who are much more familiar with the developed-country alternatives that so dominate the industrial sector of many African countries (see below, however, on the potential opportunities afforded by Chinese and Indian technology).

Decentralised institutions

In the context of road construction – which we will revisit in the cases below – Stock and de Veen (1996) have usefully drawn attention to the political economy dimension of decentralisation and compared it in this respect to centralisation. As they see it, the main task is one of enlarging and strengthening the domestic constituency in favour of labour-based methods. In centralised programs, often the only stakeholders supporting labour-based methods, other than the donors financing the program, are the small farmers in rural areas who work on the road sites, and the small-scale contractors who have little access to equipment.

In decentralised programs, however, the set of stakeholders grows to include local civil servants. These civil servants are often given to support labour-based methods because of their simplicity – they enable civil servants to manage road works that would have been managed at a higher level if carried out with equipment-based methods. In addition, decentralisation often makes it easier for the supporters of labour-based methods (the contractors, the local officials and the small farmers who work as labourers) to press their demands on government, since they may have more power at the local level and are closer to where management decisions are made (Stock and de Veen, 1996: 24)

If, therefore, in these various ways, decentralised institutions tend to favour the selection and upscaling of labour-intensive techniques, a major reason why this does not occur more often is that by comparison with Asia and Latin America, Africa tends to be a centralised state (James, 1995). As noted below, however, this general tendency does not rule out occasional examples of successful scaling up.

Aid donors

Historically, aid donors in Africa have exhibited a strong predilection for large-scale, import- and capital-intensive technologies, rather than small-scale, labour-intensive alternatives. In large part, as Tendler (1975) and others have pointed out, this is due to the perverse way in which incentives are structured in these institutions. More specifically, there are rewards to be had for administrators who are able to 'move' large amounts of aid funds. And since money is moved more easily in large units, a clear preference is created for large-scale projects, whether they are efficient or not.

The problem is only exacerbated, moreover, by the common tendency for aid projects to finance just the foreign exchange (as opposed to local) costs of projects. For then there is even more of a bias in favour of the type of technologies, as noted above, that are most effective in meeting the goal of 'moving money'.

Note, too, that a symmetrical type of process seems to be at work in the African recipient countries. Thus, according to the 'bureaucratic-man' hypothesis,[7] the officials who organise and administer aid projects are concerned to maximise the inflow of foreign exchange by supporting large-scale, import-intensive projects. In this process, the technology involved, and indeed the efficiency of the projects as a whole, is barely considered. The foreign exchange that is involved mainly serves to maximise the political power of state institutions, such as ministries, development finance institutions or the Treasury.

Consider finally – and perhaps most importantly – that aid donors in Africa are not usually given to scaling up small-scale enterprises and appropriate technologies. The chance of this happening, however, is increased by the enhanced presence of China and India in the process, since these countries are already more familiar with these technologies than is usually the case in the West. Indeed, in the 1970s, two of the most successful public enterprises in Tanzania, involving textiles and farm implements, were built with bilateral aid from China (see more below on this).

NGOs

International non-governmental organisations (NGOs) are sometimes mandated to promote appropriate technology in developing countries. I am thinking here, for example, of two such institutions in the USA and the UK, originally called Appropriate Technology International and The Intermediate Technology Development Group, respectively (their names were subsequently changed). Freed from the restrictions imposed by the profit motive, institutions such as these have sometimes seen fit to design technologies specifically for small-scale enterprises in Africa (something that, as noted below, would not be viewed as worthwhile by the typical large-scale multinational firm).

On rare occasions, NGOs that specialise in appropriate technology have even undertaken to scale up their successful innovations at the micro level to the regional or national levels (see cases below). Such examples, however, took place mainly in the 1980s, and there is a dearth of information as to whether there have been any more recent attempts. To this extent, there is a clear need for further research on the topic.

International organisations

Of the various international institutions, probably the one with the closest connection to appropriate technology is (or was) the ILO. After all, as its name suggests, this organisation is principally concerned with issues of employment and unemployment. And since technology is a major determinant of the numbers employed in a particular firm or farm, much research in the ILO has been conducted on the feasibility and efficiency of labour-intensive technology in developing countries (Bhalla, 1975).

Beginning in 1968, such research was largely financed by Swedish funds administered under the aegis of the so-called World Employment Programme (which was terminated at some time in the 1990s). The programme established, over a wide range of manufacturing industries, that the choice of labour-intensive technology would not only increase employment, but also improve efficiency, relative to the large-scale, capital-intensive alternatives that were normally selected.[8] Rarely, however, was an attempt made to follow up on the cases where appropriate technology had actually been successfully used at the micro level.[9] Apart from a few notable exceptions, few efforts were made to consider what would have been involved in scaling up what were often pilot projects to other regions and sub-regions.

In two cases, though, of rural access roads in Kenya and Botswana, scaling up was attempted and favourable results were achieved. These cases are discussed below, with the benefit of insights derived from the general literature on scaling up.

Multinationals

In a controversial volume published in 2004, C.K. Prahalad argued that there was a 'fortune' to be made by multinationals serving the poor at the base of the pyramid (BoP) in developing countries. However, because this is not at all the type of market usually served by these firms, Prahalad's argument has been seriously questioned.

Probably the most trenchant criticism has come from Karnani (2007) who questions the entire premise of selling profitably to the poor. For,

> Not only is the BoP market quite small, it is unlikely to be very profitable, especially for a large company. The costs of serving the markets at the bottom of the pyramid can be very high. The poor are often geographically dispersed (except for the urban poor concentrated into slums) and culturally heterogeneous. This dispersion of the rural poor increases distribution and marketing costs and makes it difficult to exploit economies of scale.

Weak infrastructure further increases the cost of doing business. Another factor leading to high costs is the small size of each transaction.

(Karnani, 2007: 91–92)

It is partly for these reasons, Karnani believes, that so few of the examples cited in Prahalad's book actually concern multinationals – and there would have been even fewer if the latter had confined his search only to the Africa region.

Such examples do exist, however; one of them, a mobile money application in Kenya called M-Pesa, is associated with Safaricom, a subsidiary of the British multinational Vodafone. This venture has been spectacularly successful, bringing mobile banking to almost the entire Kenyan population. But it is based on a revolutionary technology, the mobile phone, and a creative application of this technology to banking services, which meets a basic need at a price even the poor in that country can afford.

Chinese and Indian institutions

If Western donors and foreign investors therefore seem unlikely to play more than a minor role in upscaling efforts in Africa, the same may not be true of aid and foreign investment from large developing countries such as China and India. In what may be described in these respects as a new international economic order, such countries have come to play a much more important role in aid, trade and investment in Africa, than they did some twenty-odd years ago.

In the case of Chinese aid, for example,

the amount of cumulative aid between 1949 and the end of 2009 was 43 billion US$, on the basis of the exchange rate prevailing in the latter year, whereas between 2010 and 2012, the amount was of the order of 15 billion US$.

(Sun, 2014)

In terms of foreign direct investments moreover, between 2003 and 2012, 'direct investment flows from China to Africa grew at an annualised compound rate of 47.8 per cent, with investment stock increasing 52.5 per cent' (World Bank, 2015: 1).

Such numbers as these are important for my purposes because China is relatively familiar with small-scale, appropriate technologies, and is thus more inclined to use these in aid and foreign investment projects than Western actors. I have already referred to two successful aid projects in Tanzania, and there are others who have described the small-scale, labour-intensive nature of Chinese private foreign investment in Africa (Shen, 2015).

Unfortunately, however, recent case studies of Indian or Chinese technology choice or upscaling are practically non-existent. There is, in consequence, a major research gap here that needs to be filled.

Informal machinery makers

As noted above, appropriate technology is mainly concerned with the bottom deciles of the income distribution. The presence of a local capital-goods sector in a developing country promotes the scaling up of such technology, because local producers (often located in the informal sector) are more likely to be able to produce the improved technology at a price which lower-income groups find affordable.[10] This is the result, largely, of differences in factor proportions: labour-intensive producers benefit relatively from the low wages that normally prevail in the informal sector of developing countries (perhaps especially in Africa). Much also depends, though, on the technological capabilities of informal sector producers, which will vary from one country to another and from one firm to another.

Lessons from scaling up

In the previous section, I looked at institutions which are, or are not, likely to form part of scaling up efforts on behalf of appropriate technology. My concern, however, was limited mainly to the behaviour of individual institutions, rather than their interactions in those efforts. Such interactions are more nearly the subject of the general scaling-up literature, which by now is quite substantial. Indeed, it would be impractical (and arguably undesirable) to survey all of it here.

Rather, I focus on the main lessons from that literature, which can then, in the next section, hopefully help us to understand the success of the four cases that will be discussed there. Hartmann and Linn (2007) and Chandy et al. (2013) have helpfully suggested five lessons 'that emerge for scaling up most development interventions'. These are listed in Table 2.1. Note that not all of them will appear in all the cases, either because they are not relevant there, or because suitable information is not available.

Let us first consider these entries from the point of view of the analysis of institutions in the first section of the paper. We noted there, for example, that there were several prominent institutions that tended to be opposed to appropriate technology, and especially to projects that seek to scale up this form of technology. From the vantage point of the entries about politics in Table 2.1, this raises the key question of how the proponents of labour-intensive technology will gain the support of, or win over, the opposing political forces. What political alliances are possible? Who will adopt the role of long-term advocate, and which agency of the state will support the scaling-up process and provide long-term commitment to it? (The focus on the time period is important given the emphasis of the scaling-up literature on this factor.) Such support is likely to be especially necessary as the scaling-up process proceeds and new sources of opposition to it are likely to arise.[11] Not the least of the problems is that the institutions that were described above as being likely to support upscaling of labour-intensive technology, such as NGOs and decentralised local institutions,

Table 2.1 Five lessons for scaling up

Topic	
Leadership and values	'More than anything else, scaling up is about political and organisational leadership and values. If leaders don't drive the process of scaling up, if institutions don't embody a clear set of values that empower managers and staff to continuously challenge themselves to scale up, and if individuals within institutions don't have any incentives to push themselves and others to scale up successful interventions, then the current pattern of pervasive "short-termism" and fragmentation will continue to characterise national policies and programs as well as, the policies and approaches of donors' (Hartmann and Linn, 2007: 4).
Political constituencies	'Scaling up requires political commitment – creating political space is a long-term process that must be started early on in the scaling-up journey. It requires advocacy and the legitimization of the programs. This goes beyond simply informing decisionmakers about the program. It requires creating constituencies and mobilizing stakeholders who are willing to place the expanded programs on their political platforms – Advocacy, political engagement, leadership formation, and participation in the political process need to be integral parts of programs hoping to become larger and be politically sustained' (Hartmann and Linn, 2007: 3).
Incentives and accountability	'Scaling up is a change process, but changes are often stalled by unwilling players. In social-delivery programs, these players are often public bureaucracies where inertia, combined with inadequate skills and human resources, prevents change from happening. Scaling-up processes thus need to include incentives for the key actors. One important tool for creating incentives is to plan for incremental steps with early results, rather than building the perfect program to be rolled out after a long preparation time without intermediate results' (Hartmann and Linn, 2007: 5).
Monitoring and evaluation	'Monitoring and evaluation will be necessary on two levels. First, for the original limited-scale or pilot operation and, second, during the scaling-up process. The successful scaling up of the BRAC operation in Bangladesh depended crucially on regular feedback from monitoring and evaluation systems. This allowed the programs to be adjusted as they expanded' (Hartmann and Linn, 2007: 5)
Orderly and gradual process	'An orderly and gradual process, careful logistical planning, a clear definition of partners' roles, and good communication are important ingredients to scale up development interventions. Scaling up is a complex and long-term challenge' (Hartmann and Linn, 2007: 5).

tend generally to lack much in the way of political power (especially in the highly centralised African state).[12]

Consider also that the importance of the entries in the table is likely to vary according to the type of project being undertaken. For example, projects involving the scaling up of road works may tend to be more political and more time-intensive than projects which are relatively market-orientated. This proposition is illustrated in the next section, where the first two cases are from government road construction, as opposed to the third and fourth, which are drawn, respectively, from edible oil processing and improved charcoal stoves, though all the cases share the common feature that they were able to successfully upgrade a small, initial effort to the regional or national levels.

Case studies

I now turn to examine the ability of the two previous sections – on institutions and lessons of scaling up – to throw light on four case studies undertaken in Kenya, Botswana and Tanzania. What is examined, in particular, is whether the institutions described in the first section as being relevant are also those that played a predictable role in the cases concerned, and whether and to what degree the processes adopted in those cases conform to the lessons summarised in Table 2.1. Note that not all the entries in that table appear in every case study, either because they do not appear to have been relevant, or because not enough information on them was actually collected.

The Roads Programmes in Kenya[13]

Research on labour-intensive methods of road construction on rural access roads in Kenya was undertaken initially both by the World Bank and the ILO (as part of the World Employment Programme). The original goal was to establish whether such methods were feasible, and if so, whether they were also efficient (this provided the underlying rationale of the project). Once these characteristics had been demonstrated, the scaling-up project fell into the hands of the engineers at the Technology and Employment Branch of the ILO, whose task (along with other actors mentioned below) was to apply the chosen methods to rural access roads in the country. That the task was successful can be gauged from the finding that by the end of the 1980s 'Over 5,000 km of rural access road have been constructed and over 80,000 years of employment created' (McCutcheon, 1990: 115). Nor, even by then, was the scaling up of labour-intensive methods to the national level confined only to Kenya. Apart from the case of Botswana to be discussed below, national level cases existed also in Lesotho, Malawi and Mozambique (McCutcheon, 1990).

Apart, therefore, from providing the underlying vision for the project, the World Bank and the ILO also seem to have played what the scaling-up literature refers to as 'champions' in the scaling-up process. That is,

> there has to be a leader or champion. All successful programs that have expanded from small beginnings have benefited from charismatic leaders who are endowed with a vision, are persistent in their efforts, are often well connected to major stakeholders and constituencies, and have the ability to command respect and guide people.
>
> (Hartmann and Linn, 2007: 3)

Other sources of support for the roads project came from unexpected quarters. In the first place, the donors to the project had to overcome resistance to the initial pilot project from within the Ministry of Works. This was unexpected because Western aid donors, as noted above, tend to display a pronounced preference in favour of just the opposite form of technology. No less surprising was the behaviour of the Kenyan government itself, which made the type of long-term commitment to the project that is regarded as so telling in the scaling-up literature.

Evidence of such commitment is provided first by the finding that the Kenyan government has continued to allocate funds for the programme. In addition, the government itself (i.e., without donor support) began to finance maintenance (which had not initially been considered). Equally, in 1987 it formally initiated the Minor Roads Programme, which was committed to the maintenance of the 5,000 kilometres of access road and the upgrading and maintenance of 4,500 kilometres of gazetted road. It has also begun to use labour-intensive maintenance on the major road network (McCutcheon, 2008: 14).

Just what lay behind such atypical behaviour is not entirely clear, but a number of factors may have contributed to it. First, one can point to the favourable result of the pilot project, coupled with the proselytising role of the World Bank and the ILO, acting, as noted above, as champions of labour-intensive techniques. Then too there was the fortuitous fact that since 1981 the programme has been headed by a Kenyan. For most of the 1980s, the Permanent Secretary of the Ministry was a person who had previously been a District Engineer in the RARP (Rural Access Roads Programme). His personal knowledge of, interest in and commitment to the programme was invaluable: 'at critical moments the Permanent Secretary acted in support of the programme' (McCutcheon, 2008: 14). Nor should one underestimate the elevated standing, at the time, of employment as a policy goal (as reflected, for example, in the formation of the World Employment Programme in 1968). No less important, finally, was the support provided by the RARP to agriculture and rural areas in terms of employment and food production, both of which, at the time were also serious policy goals (as they were for the Kenyan government as well). Indeed, there were 'reports that at the local level politicians would not allow government to think of stopping the programme because of its dual success: employment and usable roads' (McCutcheon, 2008: 14).

Because some of these reasons apply as much to donors as to the Kenyan state, they help to explain what the programmes most clearly revealed: namely, a close alignment of interests among the major parties, one that held up, moreover, over several decades. Indeed, by 1986, the World Bank described the programme as one of the best that aid had financed in the country up till then (McCutcheon, 2008). This achievement was made all the more remarkable by the fact that the interests of the main actors did not conform to those normally expressed in the literature. Quite the contrary, in fact (in this regard it may be worth distinguishing between the World Bank and the other donors to the project, since the Bank has, at various times and places, expressed an interest in, and a preference for, labour-intensive technology[14]).

Finally, there is one other respect in which the Kenyan experience conforms to the lessons of scaling up, as were set out in Table 2.1. What I am referring to here is the entry about the need for an 'orderly and gradual' scaling-up process. The roads project seems to have met this requirement in that:

> The programme had a very slow build up – Thus in the first 3 years output was low. This was a result of a *quite deliberate policy decision*. It was recognised that this was a totally new programme for Kenya using a technology which was not, widely understood. Time was therefore, required to modify and adapt the existing procedures and to develop a suitable training programme.
> (Edmonds and Ruud, 1984: 15, emphasis in original)

I turn next to another successful scaling up of a pilot roads construction project in Africa: namely, the Botswana Rural Roads Programme. This case will be dealt with in somewhat less detail than the Kenyan example, because the experience in Botswana was modelled in key respects on the former.

The Botswana roads project

In both Kenya and Botswana there was long-run political support for labour-intensive methods of road construction. However, in Kenya,

> this was true throughout the programme whereas in Botswana during the early years it was touch and go whether the highest levels (President and Vice-President) and some senior officials would outweigh the negative views of senior members of parliament and many district officials.
> (McCutcheon, 1990: 22)

Why these particular actors held such negative views is not known, but they may well have reflected the general hostility towards older techniques that is so common in developing countries. As might have been expected, local engineers were also opposed to the labour-intensive methods of road construction in Botswana, at least in the case of gazetted (registered) roads, in spite of pressure being brought to bear on them. Then, finally, there were a series of design problems which had to be addressed (concerning the terrain, among other issues). These, though, were incorrectly interpreted as method failures (see also below) and further strengthened the opposition to labour-intensive techniques.[15]

Arrayed against these forces, on the other hand, were a number of factors and persons in favour of such techniques. In the first place, there was a sharp difference in the policy stance adopted towards non-gazetted, as opposed to gazetted, roads. In the former case, policies of decentralisation and rural development were pursued and responsibility had been granted to the district councils 'which were autonomous bodies following under the overall jurisdiction of the Ministry of Local Government and Lands (MLGL). In 1980 a pilot project of labour-intensive "district road" construction and maintenance was initiated in the Central District' (McCutcheon, 2008: 15–16).

Then, too, there was the crucial fact that the President had been the Minister of Finance and Development Planning at the time of the approval of the MLGL project. 'His personal support gave us [the ILO] the breathing space to improve the design – and then reconstruct the offending roads. Gradually the construction of acceptable roads and number of people employed overcame the anti-lobby and in 1983 the Ministry decided to expand the pilot project into a national programme' (McCutcheon, 2008: 16).

Finally, the influential role of the ILO needs to be acknowledged. For not only did this institution act as technical adviser, with the aim of replicating the RARP in Kenya, but it also helped to coordinate the relationships between the main actors. By 1990, more than 2,000 kilometres of road had been upgraded, and on an annual basis more than 3,000 people employed. The government itself ultimately pronounced the scaling up endeavour a success (as did a number of independent reviews).

Edible oil processing in Tanzania[16]

The two cases just described dealt essentially with government institutions and their interactions with aid donors in the public supply of infrastructural services. The next two cases, by contrast, deal more with institutions and individuals that were described above as being generally in favour of scaling up labour-intensive technologies, such as NGOs, informal capital-goods producers and low-income consumers. To this extent, institutions of the state, and especially those that are opposed to this form of scaling up, play a less significant direct role in the process (though they may nevertheless exert an important indirect role, such as through price setting and market regulation). Concomitantly, these next two cases focus more on the market as an upscaling mechanism.

This section deals with the successful scaling up of a technology for the extraction of edible oil in Tanzania. The technology, known as the ram press, 'is a small-scale, manual technology for edible oil extraction first disseminated in the Arusha region of Tanzania in 1986 through a project implemented by the Lutheran Diocese of Arusha' (Hyman, 1993: 429). Contributions to the design were also made, among others, by engineers at what was then called Appropriate Technology International in Washington, DC.

Let me first recount the respects in which the ram press may be described as appropriate. In particular,

> [the] ram press is inexpensive and can be manufactured and repaired in informal sector, rural workshops using labour-intensive methods. No specially imported components are needed and spare parts can be made locally. It is portable, durable, and easy to maintain. The ram press was designed for soft-shelled varieties of sunflower seeds with a high oil content – In contrast with expellers, no diesel fuel or electricity is used by the ram press.

Furthermore, the product is cold-pressed oil, which has a longer shelf life and tastes better than oil produced in a motorized expeller.

(Hyman, 1993: 429)

Other dimensions of appropriateness also warrant mention. 'Service pressing', for example, is a popular means of spreading the technology to local farmers who do not actually own it. On average, twelve other households benefitted from this process for each producer, and included both men and women. Then, in terms of employment, mention should be made of the fact that, on average, three young unskilled labourers per manufacturer were hired on a part-time basis (thereby limiting interference with the worker's scholastic obligations). There was also talk of gains on the consumption side as production increased to meet a demand for seed oil that may not otherwise have been entirely met (aided, in addition, by price reductions of the product).

By 1992, almost 800 ram presses had been manufactured in six regions of the country. No fewer than twenty-four local organisations had been involved in the process, including several that had given priority to assisting women's groups.

The wide dissemination thus achieved seems to have been due to a variety of factors, not the least of which was the interactive learning process that accompanied successive modifications of the technology. That is, far from being in any sense a 'blueprint' for progress, the process of dissemination included a key interactive learning mechanism, through which observed weaknesses in the technology were modified in subsequent vintages.

Moreover, and relatedly, the efficiency of the dissemination process was due in part to the success with which 'user needs were understood' and satisfied. For, although this seems like an almost self-evident requirement of the process, it is remarkable how few of the market-orientated approaches to scaling up actually meet the expressed needs of buyers in an affordable way (James, 1989, 1995). This, however, is precisely what seems to have occurred in Tanzania, where 'the ability to supply a new technology at a low price proved to be the key to widespread adoption' (Hyman, 1993: 441). But this favourable price was not achieved in a vacuum. It was due, for one thing, to the interest in the proven product, shown by numerous producers, instead of just one. 'Faster progress can be made by working with multiple manufacturers, because competition can keep prices down, stimulate design innovations, and encourage better quality control. With decentralised workshops, repair facilities will be located closer to the buyers'. (Hyman, 1993: 441).

Then again, there were suitable choices made by those in charge of the project – those who were responsible for guiding the market mechanism. Thus, 'the ram press projects in Tanzania were able to achieve a significant impact because they concentrated on commercialization of a single technology and adopted an integrated approach of providing assistance. This assistance included access to equipment, extension services, credit, and training of manufacturers and users' (Hyman, 1993: 441).

If, therefore, the upscaling of the ram press involved very different institutions from those that were dominant in the roads projects, there is one respect in which the cases share a lesson from the scaling-up literature. It has to do with the speed of the process, and in this regard, I have already noted that the roads projects saw gradual and orderly progress. The case of the ram press, as well, 'indicates the importance of a sustained, gradual approach to technology transfer with a time horizon of five years or more. An active role in the initial dissemination is needed to achieve widespread impacts' (Hyman, 1993: 442).

Consider, finally, the key question of whether the Tanzanian experience can be replicated in other African countries. Certainly, pilot ram press projects were being implemented even by 1990 in several such countries, but whether or not they have been scaled up successfully remains unknown, as far as I can tell. What is clear is that success elsewhere cannot, by any means, be taken for granted. For what is known is that even revolutionary and highly popular new products, such as the M-Pesa mobile money scheme in Kenya, have proven difficult to replicate in other African countries, despite their obvious benefits.

Improved charcoal stoves in Kenya[17]

This last case, of improved charcoal stoves in Kenya, has more affinity with the ram press than it does with the road projects. Like the edible oil case, it involves many institutions that were described as being conducive to scaling up small-scale projects. Moreover, it relies relatively heavily on the market mechanism. And because the history of improved cooking stoves in Africa as a whole is far from being an especially favourable one, the relative success of the Kenyan case contains much that is relevant from a scaling-up perspective.

The improved technology in question, known as the 'bell-bottom jiko', was based on the design of the traditional stove in Kenya and the addition of a ceramic liner, which, in turn, was inspired by a Thai design (whose features were investigated in detail during a study tour undertaken by a group of Kenyans to Thailand). The trip was financed by the USAID, operating through The Kenya Renewable Energy Development Project (KREDP), one of whose activities was the support of liner stoves. The design of these stoves was completed in 1983 and revised again in 1984. The subsequent rate of dissemination was quite rapid: by 1986, 125,000 ceramic lined charcoal stoves had been adopted in the country.

The improved technology was appropriate in a number of respects, the first of them being that the involvement of informal sector producers contributed to increased employment. The second respect is that the improved technology saved a considerable amount of fuel compared to the traditional model. Then again, there were large monetary gains to the consumer. Indeed, 'perhaps the key element in the success of the ceramic jiko is the irresistible cash savings it can offer to any household that can afford the modest cost of 65 to 100 shillings £2.60 to £4' (Harrison, 1987: 41).

One of the main lessons to be drawn from this experience has to do with the nature of the learning process that was adopted. For one thing, it involved upgrading of the traditional stove; that is, a model that is based on the underlying assumption that there are intrinsically favourable elements to be retained from the traditional product. For example, the new ceramic stoves did not emerge fully blown, but were rather the culmination of many successive trials and modifications to the original jiko stove, undertaken by different engineers over the years (some of whom worked for what at one time was called Appropriate Technology International, an American NGO). Also part of the learning process was the willingness of Kenyans (as noted above) to study the idea for a new model stove in another, very different, developing country, Thailand.

> There was also learning on the demand or consumption side. At one point, for example, the development project embarked, just as any commercial company would, on widespread consumer testing. It used the umbrella group Kenya Energy Non-Governmental Organisations to give free stoves to 600 households all over the country, asking only for their comments in return. The trials led to important changes.
>
> (Harrison, 1987: 41)

Of no less importance was the role played by the informal capital-goods sector; according to Hyman (1987: 385),

> the informal sector had a major role in production and distribution of the stoves. The informal sector has the ability to adapt rapidly to new designs, if they are appropriate – Reliance on existing artisans avoids the need to establish a whole new, infrastructure and completely train inexperienced workers.

The prevalence, in this case, of NGOs, informal capital-goods producers and local engineers, conforms to what one would have expected from the institutional analysis of scaling up conducted in the first section above. But the successful performance of these institutions was far from a simple matter, primarily because of the intensive learning processes that took place on both the supply and demand sides of the market.

Conclusions

This chapter has investigated the important question of why successful small-scale, labour-intensive projects in Africa are so rarely scaled up to the regional or national levels, where they impact large numbers of people. But the chapter has also sought to explain certain exceptional case studies in the region that do not conform to the usual pattern.

As a first step in the argument, I began with a discussion of how the features of appropriate technology influence the type of institutions that are likely to help or hinder the scaling-up process. For example, appropriate technology is closely related to improving low incomes, rural development and job creation, and institutions which are also concerned to promote these goals will tend to have a positive influence on scaling up as well. The converse is true of institutions that neglect or negatively influence such goals.

Then, in an effort to further enrich the discussion and prepare to understand the successful case studies, I presented some lessons from the general scaling-up literature, which has grown quite rapidly in recent years. Though not all these lessons are relevant to all the case studies, they do nevertheless introduce some important political and sociological variables into the discussion (such as, for example, the degree of political support for, and commitment to, the scaling-up process by different individuals and institutions).

Indeed, these factors turned out to be decisive in the first two cases of scaling up labour-intensive road construction in Kenya and Botswana, though the alignment of political forces took longer to forge in the latter as opposed to the former country. In terms of the discussion above, these two cases are remarkable in that Western donors, and certain important government institutions, displayed a preference for the labour-intensive technique as opposed to their usual predilection for the opposite type of capital-intensive technology.

As possible explanations for these unusual outcomes, I suggested first that labour-intensive techniques promoted rural development and employment, which were both important policy goals at the time, for donors and recipient governments alike. Then there was the fortuitous fact that in both countries, a senior figure in the relevant Ministry supported the scaling-up programme and his support for labour-intensive methods was crucial at key phases of the project.

The third and fourth cases differ from the first two in that they are more concerned with the market as a mechanism for producing and distributing the two new technologies, the ram press in Tanzania and the ceramic charcoal stove in Kenya. To this extent, state institutions played a lesser role than they did in the first two cases. All the same, successful market-oriented scaling up proved to be quite a complex affair. In both cases, for example, there was a good deal of interactive learning, as the respective technologies underwent continuous changes and improvements (indeed, in Kenya, learning was undertaken so intensively that a study tour took place to investigate a technology variant in another developing country in Asia). So too were there deliberate and intensive efforts to understand user needs, a crucial endeavour whose salience was emphasised above. And finally, what should not go unmentioned is the role of the informal capital-goods sector in both countries, which involved producing the appropriate technology more cheaply than previous variants, and was more appropriate in other key respects as well.

Notes

1 This was noted by Carr as far back as 1985.
2 See, for example, the discussion in White (1985).
3 See, for example, Hartmann and Linn (2008) and Chandy et al. (2013).
4 Some of the entries in Table 2.1 make reference to these considerations.

5 For a discussion of the goals of institutions, as they relate to appropriate technology, see Stewart (1987).
6 See James (1995).
7 See Williams (1975) and James (1995).
8 See ILO (1992).
9 See ILO (1992).
10 See Pack (1981) for a discussion of how an indigenous capital-goods sector promotes appropriate technology in developing countries.
11 This also is emphasised in the scaling-up literature.
12 But note the role played by informal capital-goods producers in the third and fourth cases below.
13 The description of the roads projects in Kenya and Botswana draws on McCutcheon (2008).
14 There may also be differences in this respect between bilateral donors. Note that UNIDO at one point produced a series of guides to appropriate technology.
15 The design problems, however, were later solved.
16 The case description below draws on Hyman (1993).
17 This case follows Hyman (1987).

References

Bhalla, A. (ed.) (1975). *Technology and Employment in Industry: A Case-Study Approach*, Geneva: ILO World Employment Programme Research.

Carr, M. (1985). *The AT Reader: Theory and Practice in Appropriate Technology*, London: Intermediate Technology Publications.

Chandy, L., Hosono, A., Karas, H. and Linn, J. (2013). *Getting to Scale: How to Bring Development Solutions to Millions of Poor People*, Washington, DC: Brookings Institution Press.

Edmonds, G. and Ruud, O. (1984). *Labour-Based Construction and Maintenance: Some Indicators of Viability*, Geneva: ILO World Employment Programme Research.

Harrison, P. (1987). 28 May.

Hartmann, A. and Linn, J. (2007). *Scaling Up: A Path to Effective Development*, Washington, DC: Brookings.

Hyman, E. (1987). The strategy of production and distribution of improved charcoal stoves in Kenya, *World Development*, 15,3:375–386. doi.org/10.1016/0305-750x(87)90019-2.

Hyman, E. (1993). Production of edible oils for the masses by the masses, *World Development*, 2,13:429–433. doi.org/10.1016/0305-750x(93)90115-3.

ILO (1992). *The Impact of World Employment Programme Research on Technology*, Geneva: ILO.

James, J. (1989). *Upgrading Traditional Rural Technologies*, London: Macmillan.

James, J. (1995). *The State, Technology and Industrialization in Africa*, London: Macmillan.

Jéquier, N. and Hu, Y. (1989). *Banking and the Promotion of Technical Development*, London: Macmillan.

Karnani, A. (2007). The mirage of marketing to the bottom of the pyramid, *California Management Review*, 49,4:90–111. doi/pdf/10.2307/41166407.

McCutcheon, R. (1990). Labour-intensive road construction in Africa, *Habitat International*, 13,4:109–123. doi.org/10.1016/0197-3975(89)90042-8.

McCutcheon, R. (2008). Labour-Intensive Construction and Maintenance in Sub-Saharan Africa, paper for discussion at the World Bank, 7 April.

Pack, H. (1981). Fostering the capital-goods sector in LDCs, *World Development*, 9,3:227–250. doi.org/10.1016/0305-750x(81)90028-0.

Prahalad, C. K. (2004). *The Fortune at the Bottom of the Pyramid*, Philadelphia: Wharton Publishing, University of Pennsylvania Press.

Shen, X. (2015). Private Chinese investment in Africa: Myths and realities, *Development Policy Review*, 33,1:83–106. doi.org/10.1111/dpr/12093.

Stewart, F. (ed.) (1987). *Macro-Policies for Appropriate Technology*, Boulder: Westview.

Stock, E. and de Veen, J. (1996). *Expanding Labour-Based Methods for Road Works in Africa*, Washington, DC: World Bank Technical Paper no. 347.

Sun, Y. (2014). *China's Aid to Africa: Monster or Messiah?*, Washington, DC: Brookings, East Asian Commentary.

Tendler, J. (1975). *Inside Foreign Aid*, Baltimore: Johns Hopkins University Press.

White, J. (1985). External development finance and the choice of technology, in J. James and S. Watanabe (eds.) *Technology, Institutions and Government*, London: Macmillan.

Williams, D. (1975). National Planning and the Choice of Technology: The Case of Textiles in Tanzania, Economic Research Bureau Paper no. 75, 12, Dar es Salaam.

World Bank (2015). *Manufacturing FDI in Sub-Saharan Africa*, Washington, DC: World Bank.

3   The changing institutional environment for technology in Africa

After seeing a period of intense interest from the 1970s to the early 1990s, research on appropriate technology in Africa (and, as far as I can tell, in other regions as well) has declined precipitously over the past twenty years or so.[1] I focus on Africa partly because the steepness of the decline there may have been more pronounced than in other regions, but also because the consequences tend to be more severe (and the need for a revival of such research, to that extent, more pronounced). I argue in this regard, for example, that Africa has been especially heavily influenced by the increased sphere of influence of China, India and certain other developing countries in R&D, foreign investment, trade and aid. Yet, remarkably, there is almost no research on the choice of technique by actors from these countries: whether, for example, their choices are more or less labour-intensive and efficient than local or developed country firms in the same branch of manufacturing. Nor, equally, is there much work on whether Chinese and Indian products in Africa are appropriate for the needs of poor consumers in the region. Some appear to be, but others are instead of unacceptably low quality.[2] (Appropriate products were first analysed in a developing country context by Stewart [1977] and James and Stewart [1981] using Lancaster's product characteristics approach to demand theory.)

Whether the decline in appropriate technology matters also depends on why it occurred and what had been achieved in the earlier period of rapid growth of the topic. It would be one thing, for example, if a wide variety of labour-intensive and efficient technologies had been identified and widely diffused across the manufacturing and other sectors (such as infrastructure) in sub-Saharan Africa. But while this has happened in a few exceptional cases (see above), on the whole the appropriate technology movement has been successful only in *identifying* appropriate technology. In other words, the available research is much more about the potential of this technology than it is about the actual diffusion of it, and there remains much to be done in understanding how the former gets converted into the latter.[3]

Why, then, did research on appropriate technology in Africa go through such an intense dry period over the past twenty-odd years? It is true that the development literature has had its share of fads and fashions over the years, with surplus labour and basic needs being just two of them. Appropriate technology research may have suffered from something of this same tendency, especially if the researchers involved had inclined to the view that showing the potential of the concept was the main goal.

My view, however, is that the story of appropriate technology research has more to do with changes in the institutional environment that bore on the support for, and financing of, this type of research. I am referring here mainly to the World Employment Programme run by the ILO since 1969, but there are other institutions as well. Until it completed its research on technology and employment some time in the 1990s, this programme may well have financed much of the work on appropriate technology in developing countries in general and Africa in particular. I shall be drawing on some of the results of this research in the sections below.[4]

In seeking to decide whether the decline of research in appropriate technology matters, I also need to refer to the ever more acute problems of supply and demand in the labour market for manufactured goods in sub-Saharan Africa. On the supply side, there are problems of rapid population growth and excess labour supply (which are only exacerbated by the displacement of workers caused by Chinese imports). On the demand side, too, there are reasons for concern caused by ongoing advances in automation and labour-saving technological progress. These problems, I should emphasise, come on top of already high rates of un- and under-employment in much of the region (see ILO Yearbooks for the latest estimates of these amounts).

In the face of these worsening rates, moreover, there may be rising levels of frustration and disappointment, which ultimately bear negatively on the incidence of crime (as measured, say, by homicides) and occurring, among other ways, through gang activities. It is not difficult to imagine that such a mechanism, on a large scale, might imperil the entire process of industrialisation in a fragile low-income African country. If it is widely spread, however, appropriate technology has the ability to reduce the likelihood of such a violent outcome.

Potential vs actual use of appropriate technology

It is useful for this and subsequent sections to provide a simple denotation of appropriate technology as the concept is used in much of the development economics literature.[5] In Figure 3.1, for example, the curve of equal output (the isoquant) denotes all the possible combinations of capital and labour, and the one that is chosen will depend on the relative price of capital and labour. In the rich countries, where the price of the former input is low relative to the latter, the lowest cost of producing the given output occurs at point A. In poor countries, on the other hand, where labour is cheap relative to capital, the point of tangency takes place at B. The additional employment created in the latter countries is equal to L_1L_2. Since it is also more efficient at the prevailing factor prices, the developing country choice can be described as appropriate.[6] (Though, as argued below, this analytical framework contains several important limitations.)

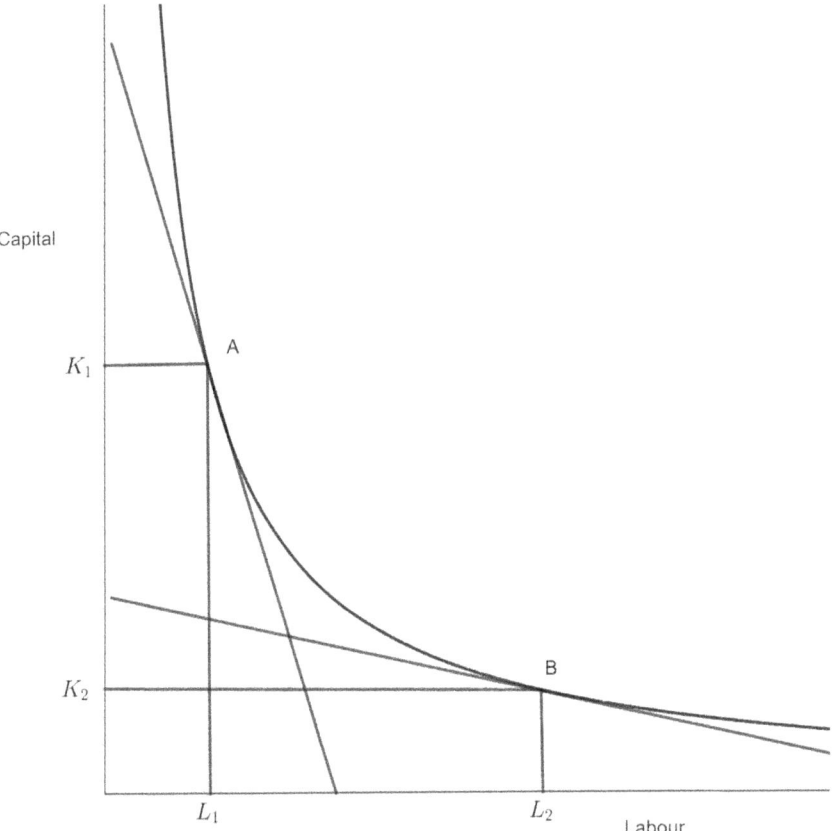

Figure 3.1 The choice of techniques in Africa

Probably the main contribution of the appropriate technology literature has been to demonstrate that such techniques as this one actually exist, across a wide swathe of manufacturing industries in sub-Saharan Africa and elsewhere in the developing countries. Some of the most influential early publications in this regard were written by Stewart (1977) and edited by Bhalla (1975). A list of industries in which the relevant research has been conducted was compiled somewhat later by Pack (1982) and is shown in Table 3.1.

What the Table shows for each of nine industries is that a labour-intensive technology co-exists with an alternative capital-intensive version with a higher capital to labour ratio. Other data, moreover, confirm that in each case the former alternative is more efficient than the latter.[7] Thus summarised, the literature had demonstrated the potential of appropriate technology: that is, its ability to contribute jointly to the goals of employment creation and economic efficiency. In some cases, it was assumed that appropriate technology would automatically

Table 3.1 Labour– vs capital-intensive techniques in African manufacturing

Sector	Annual output of plant	Capital-intensive technology			Appropriate technology	
		Investment in thousands of dollars	Number of workers	Thousands of dollars per worker	Investment in thousands of dollars	Number of workers
Shoes	30,000 pair	334	155	2.2	165	218
Cotton weaving	40,00,000 sq. yards	9,779	260	37.6	4,715	544
Cotton spinning	2,000 tons	1,440	98	14.7	480	240
Brickmaking	16,000,000 bricks	3,437	75	45.8	796	238
Maize milling	36,000 tons	613	63	9.7	219	96
Sugar processing	50,000 tons	6,386	1,030	6.2	3,882	4,986
Beer brewing	200,000	4,512	246	18.3	2,809	233
Leather processing	600,000 hides	6,692	185	36.2	4,832	311
Fertiliser	528,000 tons of urea	34,132	248	137.6	29,597	242

Source: Pack (1982, Table 1)

be adopted by sectors concerned to promote at least one of these goals and possibly both of them (in the case, for example, of employment in state-owned enterprises).

It soon became apparent, though, that firms – both private and public – tended to choose the capital-intensive, inefficient options even when the alternative, appropriate technology was available (as described in Table 3.1).[8] And even when a single sector did happen to choose the latter, it rarely got adopted by the wider community. That is to say, the potential of appropriate technology has generally not been realised at more aggregative levels of economic activity. In 1975, in an observation that is still relevant today, Carr (1985: 369) put it this way:

> Successfully introducing an improved plough, mill, oil-press or water or sanitation system to a few villages, or introducing a new construction or industrial process to a few co-operatives or entrepreneurs is all very well, what is needed, however, is that such products and processes should be adopted, and come into use, in thousands, if not millions, of villages throughout the Third World.

Some years later, in a valuable simulation exercise, Pack (1982) attempted to quantify the macro potential of appropriate technology across the same sectors shown in Table 3.1 for a 'typical' African country. In particular, for each such sector, he estimates the disparities in employment, value added and capital per worker that would have resulted from an investment of $100 million in the two types of techniques shown in that table. The results are presented in Table 3.2.

Thus, assuming that at the time the entire $100 million was invested in the appropriate technology in each sector, the aggregate employment gain would have been equal to 180,661 (i.e., 238, 678–58, 017).

There would also have been a gain in the value added of $260 million per annum.[9] And since we know that in most manufacturing sectors in Africa, the inappropriate technology tends to be chosen, these two aggregate gains would at least seem to approximate what then would have accompanied a widespread shift to appropriate versions. Nowadays, of course, with continual labour-saving technical progress in the developed countries, the potential gains in employment may be even greater (though see below on the countervailing Chinese technical contribution in Africa).

Table 3.2 Employment and output implications of two technologies at the macro level

Technique	Value added ($ million per annum)	Employment (all workers)	Capital-labour ratio ($ per worker)
Appropriate	624	238,678	3,771
Capital-intensive	364	58,017	15,513

Source: Pack (1982, table 3)

Table 3.3 Labour intensity, selected branches of manufacturing, South Africa (descending order)

Branch	Labour-capital ratio*
Clothing	62.64
Furniture	31.21
Footwear	27.53
Leather	18.47
Metal excl. machinery	11.86
Textiles	11.52
Wood & wood products	11.29
Plastic	11.23
Rubber products	4.84
Food	4.29
Beverages	2.93
Tobacco	2.73
Glass & glass products	2.29
Paper & paper products	1.71

* Note: mean of 2006–2009, current prices.

Source: Tregenna (2012)

Note further that while the employment gains from an altered choice of technology may thus be considerable, the scope could be increased even more by including the choice of sector as a variable. As shown in Table 3.3 for South Africa, there is wide variation in the capital/labour ratios of manufacturing sectors and a focus on say, garments, rather than food, would have an appreciable effect on employment, even without an altered choice of technology within each of them.

It would thus aid the planning effort if the two choices – of sectors and technologies – could be made simultaneously rather than in isolation, given the observed relationship between them. In fact, one could then conceive of a two-by-two matrix of possibilities. At the two extremes would be i) relatively labour-intensive sectors and employment-intensive technologies within these sectors, and ii) relatively capital-intensive sectors with inappropriate choices made within them. It seems plausible to suggest, moreover, that this distinction can fruitfully be applied to contrast the early East Asian manufacturing experience on the one hand with sub-Saharan Africa on the other. Certainly, light manufacture such as garments were encouraged (by means for example of export processing zones) in the former region, on the basis of labour-intensive adaptations of Western technologies (Amsden, 1977; Fransman, 1984), as opposed to the latter region, which has not generally emphasised either those types of goods or low-cost ways of producing them (Stewart et al., 1992).

However, neither the comparison between the two regions nor the study of their individual experiences has been exhaustive. Because there is still much that needs to be understood, it seems as if the saturation of the field cannot readily be used to explain the relatively sudden and near total decline of appropriate technology research in Africa. Though there have indeed been many explanations for the typical choice of technology in sub-Saharan Africa, policies have tended to ignore them or they (the policies) were inadequate in the first place. Nor has enough attention been paid to the few 'success' stories, where appropriate technology has been widely spread in an African country. As shown in Chapter 2, labour-intensive road construction on minor roads in Kenya (ILO, 1992) is one such exception and small-scale edible oil processing in Tanzania is another (Hyman, 1993; see also Linn et al., 2010).

Why appropriate technology research diminished

I have argued that many demonstrations of the existence of appropriate technology in manufacturing and certain parts of infrastructure may have led to some saturation of the literature during and after (say) the mid-1990s. In this section, however, I suggest that there were other, possibly more important, reasons for the diminution in the literature on the subject.

These had partly to do with international or internationally oriented institutions that ceased or diminished their research on appropriate technology near the end of the period in which this topic was still in favour. I am thinking here primarily[10] of the World Employment Programme (WEP) of the ILO, which, with the help of the Swedish government, financed research on the choice of technology and other related issues. Since its inception in 1969, this programme had, by the early 1990s, published a wide range of books, articles and working papers on various aspects of technology and employment (ILO, 1992). Apart from its specific support for appropriate technology, WEP research may also have helped to raise awareness of the employment problems in policy-making by African governments (see also below).

The decline in research noted above may also have had to do with the emergence and growth in interest in information technology, the point here *not* being that such technology is necessarily inappropriate. On the contrary, as shown in the examples in Table 3.4, these products have already been widely diffused

Table 3.4 Ownership of mobile phones at the base of the pyramid (selected African countries 2011)

Country	Per cent ownership of mobile phones
Uganda	0.47
Kenya	0.65
Ghana	0.51
Nigeria	0.61
South Africa	0.77
Botswana	0.64

Source: Research ICT Africa: survey data

among the poor, in at least the countries shown there, and since the date when the data were collected, the numbers will surely have grown even higher.

In Kenya, mobile phones serve as the basis of the well known mobile money banking service known as M-Pesa, which serves a high percentage of the population, including some of the poorest groups (James, 2016). My point is thus not that the research on information technology is inherently undesirable; the more limited point is that this research may in part have been substituted for the more conventional literature on appropriate technology. Note, in this regard, that the United Nations University set up a centre for the study of new technology in developing countries in the 1980s, in Maastricht, the Netherlands.[11]

Note further that information technology itself may be appropriate or inappropriate for developing-country use, depending on how it is actually exploited. In developed countries, use of the technology is dependent almost entirely on ownership, whereas in the developing world, much more emphasis is given to intermediaries who come between the technology and the customers. One of the most impressive examples of this type of institutional innovation occurs in Bangladesh, where, under the aegis of Grameen Telecom, villagers effectively rent phone time from selected women who purchase phones with loans from the project. As a result of such separation of use from ownership, the village phone scheme has reached tens of millions of farmers in the country. In Africa, as well, mobile phone rentals are a common sight on many city street corners.

There is, finally, the possibility that the importance of employment creation has waned as a goal of government over the past twenty years. If so, the pursuit of labour-intensive technology will not have been viewed with much enthusiasm. I cannot, however, do more at this stage than advance this as a theoretical possibility, since it is so difficult to confirm empirically. For example, while the upper echelons of the state hierarchy may favour employment creation, it is often difficult to translate this preference into reality by lower level state agents, who may possess a different ranking. In other words, it is often difficult to speak of a state's welfare function. Even if there was such a construct, moreover, state goals may be frustrated by the actions of foreign collaborators on development projects. For instance, choice of technology decisions are often made at an early stage of the project cycle by developed-country consultants, engineers and aid donors. Such actors as these often favour capital-intensive projects, partly because they are more familiar with them. It thus remains an open question whether or not employment creation has fallen in the ranking of development goals over the last twenty years.

Does it matter?

This section advances the claim that there are at least two main reasons why the virtual absence of appropriate technology research matters a great deal in contemporary sub-Saharan Africa. One has to do with the fact that present, and especially future, conditions in the labour markets of most countries in the region are such as to make it more, rather than less, pressing for this technology to be adopted on a wide scale. The second reason confronts the fact that the new international economic order poses major new questions for appropriate technology in Africa, questions having to do more specifically with the sharply increased role in the region, of developing countries such as India and China.

Conditions in the labour market

It is convenient, in explaining the argument of this section, to use elements of the well known model of industrialisation and development first set out by W.A. Lewis in 1954 and revisited by him twenty-five years later in 1979.

In Figure 3.2(a), the Lewis ideal-type, the initial equilibrium occurs at point A, where the demand curve for labour LL and the perfectly elastic supply curve intersect.[12] The parallel outward shift in the demand curve occurs because so-called neutral technological progress occurs (that is, as output expands, the capital-labour ratio stays constant). The process continues until the entire supply of labour in the informal sector is absorbed by the formal part of the economy, at which point labour ceases to be scarce (at J) and the supply curve turns upward (the so-called Lewis turning point).

In Figure 3.2(b), however, the situation changes on both the demand and supply sides of the labour market. As regards the former, the demand curve shifts non-uniformly to the right, reflecting labour-saving technical progress, with the result that less labour is absorbed than in the previous diagram (compare the distance YM in the two figures). To make matters worse, in the actual situation there is rapid population growth and a consequent *elongation* of the supply curve. In the new equilibrium, the amount of labour left in the informal sector is greater in the second diagram as compared to the first.

In his 1979 revisitation of the original model, Lewis acknowledged that the situation depicted in Figure 3.2(b) better reflected the post-war situation in developing countries than did the ideal-type representation of Figure 3.2(a). And one of the major reasons for this, he surmised, was the asymmetric shift in the labour demand curve shown in Figure 3.2(b) caused by labour-saving technologies imported from the developed countries, where labour is scarce relative to capital.

Consider, from this point of view, the case of South Africa. In particular, 'An outstanding feature of this country's production activity is its growing capital intensity over time. This is revealed in increasing capital/labour and average capital/output ratio' (McCarthy, 2005: vii). In fact, during the 1990s, South Africa experienced an extreme form of the situation depicted in Figure 3.2(b). For not only was there then an asymmetric shift of the demand curve shown in that figure, but the twisted shape of the curve was such as to cause the new equilibrium to occur at the same point as the old. This extreme case is known as 'jobless growth', in which, despite growth in output, there is no increase in employment. Indeed, South Africa actually experienced negative growth in jobs in the manufacturing sector during the 1990s (McCarthy, 2005).

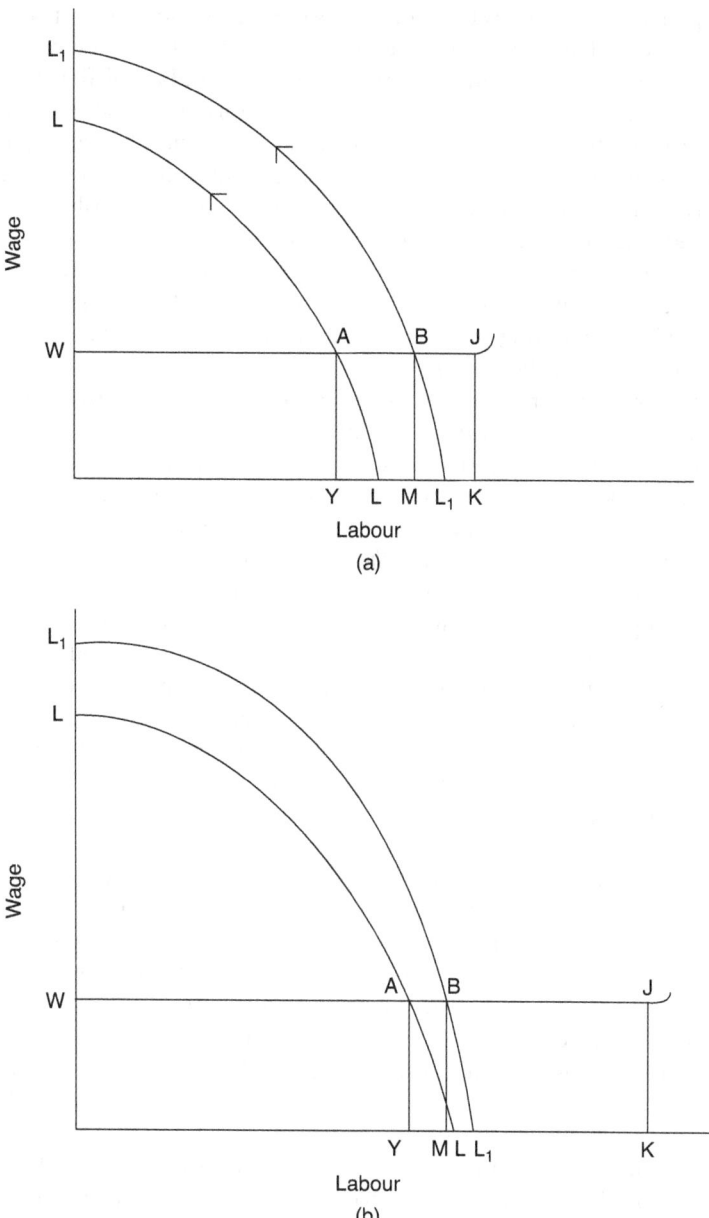

Figure 3.2 Conditions on the labour market

Source: Based on Lewis (1954) Table 3.1

Note, though, that the Lewis ideal-type model applies much better to the early experiences with industrialisation in some of the now developed countries and parts of East Asia more recently. As far as the former are concerned, a useful example comes from the early stages of the Industrial Revolution in Great Britain, where capital and labour in manufacturing grew at the same rate of 2.5 per cent. Technical change, that is to say, was neutral, according to the definition used above. In some of the European early follower countries, too, strenuous efforts were made to curtail the growth of the capital-labour ratio, as Landes (1965) has indicated. He notes, for example, that 'In Europe the follower countries made the most of their cheap manpower by building more rudimentary but less expensive equipment, buying second-hand machines whenever possible, and concentrating on the more labour-intensive branches or stages of manufacture' (Landes, 1965: 116).

Perhaps more familiar are the pervasive labour-using adaptations of imported technology in successful East Asian countries such as Korea, Japan and Taiwan. In Korea, for example, the process was so intensive that for part of the period after 1964, the capital-labour ratio in manufacturing actually fell (Ranis, 1973). In Japan, too, that ratio remained basically the same during the last ten years of the nineteenth century, '*indicating the effectiveness of capital-stretching innovations at the aggregative level*' (Ranis, 1973: 402, emphasis added).

On the supply side, too, conditions in East Asia were often historically more propitious than those prevailing currently (and expected to prevail also in the medium run) in sub-Saharan Africa, looking more like the situation depicted in Figure 3.2(a) than in Figure 3.2(b). These conditions were usually not just serendipitous: rather they were, to a large extent, the result of deliberate policy measures to bring down the rates of population growth. As Andrew Mason (2001) has put it in his detailed study of the subject:

> the governments of East Asia changed course with respect to population policy early in the post-World War II era. . . . Within a relatively brief period of time, pronatalist policies were abandoned and, over time, a wide variety of antinatalist programs and policies were adopted. Governments engaged in educational programs, increased the availability of contraceptive supplies and services, urged their citizens to adopt small family norms, and relied on incentives and disincentives to encourage couples to bear fewer children.
>
> (Mason, 2001: 8)

Although many population policies were also pursued in parts of sub-Sahara, they have enjoyed very little success. Table 3.5, for instance, shows that, at 3.1 per cent, the population there is projected to grow faster than any other region in the world.

At the projected rate, the African population will double by the year 2050. What is even more disturbing, however, is the prediction that youth un- and under-employment will have doubled five years earlier, in 2045 (African Economic Outlook, 2012). It is a matter of considerable concern because at some

Table 3.5 Average annual growth rates of population between 2015 and 2050, all regions

Region	Per cent change
World	0.92
Africa	3.1
Asia	0.57
Latin America and Caribbean	0.68

Source: UN (2015)

point, rampant youth un- and under-employment are likely to be translated – via disillusionment, frustration and anger – into violent crime such as homicides.

> A sense of injustice and anger at one's inability to earn a decent living, for example, are common causal features of the violence experienced in civil wars. They were also described . . . as being important to the gender violence that was committed in East Africa, when men's socio-economic status was . . . severely threatened.
>
> (James, 2017: 48)

If such outcomes as these are to be averted, relevant policies will need to be adopted on both the demand and supply sides of the problem. The problem with policies designed to reduce population growth, however, is that they will work, at best, only in the longer run. For even if current fertility rates can be made to fall, rapid population growth will continue well into the future. To this extent, therefore, it will have to be the demand side which bears the brunt of any countervailing policy. This means, among other things, the need to influence the choice of manufacturing sector and the choice of technology within each sector, as the East Asian experience well illustrates. The aim, as shown in previous figures, is to reduce the capital-labour ratio in manufacturing or to keep it from increasing (assuming, of course, that relatively labour-intensive techniques are also efficient, as indeed usually seems to be the case).

Yet, just at a time when these issues have become most pressing, there is, paradoxically, a near-total lack of research on them. Most notable, perhaps, is a severe lack of understanding on why only very few appropriate technologies get widely diffused across the population in African countries. This issue warrants especially close attention in view of the already substantial and growing un- and under-employment in the region and the socio-political problems likely to be wrought thereby (as noted above).

There is a need, in the first place, to identify and compare successful cases of diffusion in different countries in the region, where success is measured in terms of the spread of appropriate technology among the population (as well as effective use of that technology). There is also much to be learnt from the general 'scaling up' literature that has grown rapidly in recent years (i.e., the scaling up of a successful micro innovation to a more macro level, involving large numbers of the poor). As shown in Chapter 2, a common feature of such examples is that serious follow-up occurred following an initial demonstration (at say the pilot level) that the technology was appropriate (in a sense which should be broad enough to include information technology such as low-cost mobile phones and mobile money). There are also lessons to be drawn from cases of success outside the Africa region, the best-known and most spectacular of which is almost certainly the firm called 'Nirma' in India.

With a detergent of the same name, this firm began in the mid-1990s on a very small scale, producing labour-intensive detergents at about half the price of the then leading brand, made by Hindustan Lever. One means by which such a remarkably low price was achieved, lay in the area of product design and in particular, the design of a product for low rather than high-income consumers. Thus,

> 'Nirma' does not contain characteristics such as optical whiteners and scents, or ingredients designed to reduce the harshness of the product on hands and fabrics. Packaging, too, is an area in which there are important differences that bear on the costs of producing 'Nirma' as opposed to the competing high income products. Whereas, for example, the latter brands are packaged in heavy cardboard, on which there is also sophisticated and expensive printing, 'Nirma' is sold instead in simple plastic bags on which the printing is relatively crude and hence less costly.
>
> (James, 2000: 116)

Nor can one overemphasise the role of the choice of technology itself, which involved nothing more than hand-mixing of the ingredients in agglomerations of small workshops, without even the use of electric power. For these, and other reasons too numerous to discuss here, 'Nirma' achieved great success: indeed, by the end of the 1980s, that product 'had captured no less than 60 per cent of the entire powder detergent market . . . with regular sales to some 70 million Indian households' (James, 2000: 112).

What I conclude from this section, thus, is that because of a worsening situation in the labour market, there is every likelihood that un- and underemployment will become more serious in Africa's manufacturing sector. Such a scenario, moreover, may well be accompanied by a rise in crime and political instability, which may imperil the entire development process. Just at a time when research on appropriate technology and employment is most urgently warranted, however, it is unfortunately least in evidence. This research needs to focus heavily on how such technology, which begins on a small scale, can be replicated on a much more macro level. Given the scope of the problem, isolated examples of appropriate technology and products will not be enough to make more than a small dent in it.

Some hope arises, however, from the possibilities afforded by the new international economic order, to which I next turn.

Appropriate technology and the new international order: a research agenda

To the arguments that have just been advanced should be added the emergence in recent years of a new international economic order, which has quite radically altered the environment in which Africa currently finds itself. For, as I shall argue in this section, the new order presents possible alternatives for a relatively large shift towards a more appropriate manufacturing technology for the region. Such an opportunity, however, has been largely ignored, in part because of the absence of conventional choice of technique studies (though the theoretical case has been made in part by Kaplinsky (2011)).

What I mean by the term 'new international economic order' is the composition of global R&D, foreign direct investment, trade and aid, and the changes therein over time. In particular, that across all these fields there has been quite a sharp increase in the share of developing countries as far as Africa is concerned (most especially China and India). Consider, first, the changing composition of global R&D. Some forty years ago, only about 2 per cent of the total was concentrated in the developing countries, and of that not all was focused on the problems of those countries (Singer, 1970). However, by 2010 that number had increased to 20 per cent[13] and is now almost certainly even higher. With respect to the composition of foreign direct investment in Africa, by far the most noteworthy finding is the rise of China. For instance, between 2003 and 2012, Chinese investment in Africa grew at an annual compound rate of nearly 50 per cent and showed signs of a shift towards manufacturing industry. With respect to trade between sub-Saharan Africa and China, there has been an annual increase of 26.2 per cent between 1995 and 2010. And note, finally, that Africa became the main regional component of Chinese aid between 2010 and 2012.

The idea that these changes might have produced a set of more appropriate technologies and products in Africa is based broadly on the notion that there is a tendency for countries with similar income levels to have similar preferences and problems.[14] That is to say, innovations in China, India and elsewhere in the developing world are more likely to suit African conditions than technologies imported from the developed countries. And even if Chinese technologies are not the subject of R&D, they will nevertheless tend to be older and hence more likely to be appropriate compared to relatively recent vintages. Yet, as noted above, the appropriate technology framework has barely, if at all, been applied to test these propositions, with respect to either processes or products. Pigato and Tang (2015), for example, pointedly note that 'Very little is known about the relative factor intensity of Chinese investment in Sub-Saharan Africa and its contribution to job creation' (2015: 11).

Several authors have suggested, however, that if appropriate technology is to be found in Chinese direct investment in Africa, it is likely to be associated with private (as opposed to public) endeavours. This form of investment, we should note, has grown since 2000 at a rate that has been called nothing short of 'spectacular' (Pigato and Tang, 2015). In 2002, for example, twenty-one Chinese investment projects were privately owned; by 2013, there were 2,217 such projects out of a total of 2,282 (i.e., 53 per cent).

Not a great deal, however, is known about this form of foreign investment from China. But a case study of Tanzania suggests that it may, in some respects, be advantageous for sub-Saharan Africa. This possibility derives, in part, from the finding that 'most Chinese private firms are involved in low-tech, labor-intensive industries, such as light manufacturing and assembly . . . and many compete with domestic companies in Tanzania' (Pigato and Tang, 2015: 17). While Chinese private investment to Tanzania, therefore, tends to be concentrated in labour-intensive branches of manufacturing, such as garments and textiles, this does not tell us about the technical choices made within those branches. It is quite possible, for example, for a textile factory to use relatively capital-intensive methods of production.[15]

There remains much to be learnt, therefore, about whether an appropriate combination of sectors and technologies can be used to promote employment through Chinese private firms in African industry.

Possible future research on Chinese private firms in Africa, should seek, as a priority, to understand how they choose technology, what factor proportions are used and how these compare with other firms in the same sector. Such research also needs to take into account the fact that Chinese foreign investment often tends to displace local firms in the same sector, as also occurs with regard to imports of manufactures from that country into Africa. This again is not an area which is exactly teeming with research, but one study, of South Africa, has made some careful estimates of the magnitudes involved. These are contained in Table 3.6.

Thus, the share of Chinese goods in total South African imports between the two periods, shows mostly a rapid rate of growth. This led, in the case of clothing and footwear, to a near dominance of South African imports. More generally 'labour-intensive industries were particularly badly affected by Chinese imports

Table 3.6 Share of Chinese products in total South African imports, selected branches of manufacturing

Manufacturing branch	1995 per cent	2010 per cent
Spinning and weaving	6.4	43.5
Clothing	29.0	75.1
Leather and leather products	10.2	49.0
Footwear	35.5	76.8
Furniture	2.4	48.1
Glass and glass products	3.1	38.6
Rubber products	0.5	23.1
Household appliances	13.7	62.6
Electric lamps and lighting	9.4	59.9
Paper and paper products	0.2	8.6

Source: Edwards and Jenkins (2013: 7)

implying that the negative impact on employment was more than proportional to the output displacement' (Edwards and Jenkins, 2013: 1).

It is not inevitable, though, as Edwards and Jenkins (2013) point out, for such an increase in imports to displace local goods and jobs. For Chinese imports may take the form of intermediate inputs, which, being lower in price, would have tended to stimulate local production (on average, imports from China tend to be 50 per cent cheaper than domestic South African brands).[16]

Moreover, imports of manufactures may displace competing foreign brands rather than domestic goods. Such considerations notwithstanding, however, jobs were lost in South Africa, and plenty of them. Edwards and Jenkins (2013) put the figure at 75,000, which, in a country with such high unemployment,[17] can only be considered as very damaging, not only in economic terms, but also (and more speculatively) from a socio-political point of view.

Against such losses, however, need to be set the lower average cost of Chinese imports noted two paragraphs previously (as noted there, the price disparity is generally substantial). But quite apart from the technical difficulties involved in comparing the gains and losses of different actors, there is a further problem. That is, at least part of the explanation for the price difference between Chinese imports and local goods appears to be due to lower quality of the former. Certainly, there are widespread reports of this problem across sub-Saharan Africa, so it is unlikely to be due to just a few aberrant incidents. The policy point is that the more quality differences account for the disparities in prices between local goods and Chinese imports, the stronger is the case (with other things being equal) for imposing some form of protection against the latter, or, in the extreme case, of disallowing them altogether.

So far, however, the quality issue has been very largely ignored in development, as indeed have most topics having to do with consumer behaviour.[18] This is not altogether surprising. Some product characteristics, for example, are inherently difficult to measure, while those that are more nearly objective may need technical measurement in a laboratory. In some cases, quality differences may be slight, whereas in others they may be substantial (possibly with the imported good still being purchased, because of the price disparity noted above). A study of user perceptions of the competing products may prove useful in this regard.

What also requires original research is the last component of the new international order: namely, the effect of increased Chinese (-tied) aid on the appropriateness of industrial technology in Africa. For while it is reasonable to expect a positive influence on the historical-economic grounds noted above, it is by no means a foregone conclusion. It is not obvious, for example, that the Chinese invariably favour labour-intensive, efficient technologies. They may be more concerned with modernisation and large-scale factories, which favour capital intensity in the choice of technology. Technological capabilities in that country are by now sufficiently advanced to produce such technology in a way that they may not have been in, say, the 1970s. Much would seem to depend on who in the aid project is actually responsible for the choice of technique, whether they be engineers, economists or bureaucrats.

A lot may also hinge on the stage of the project when the decision on technology is made. Research by Jéquier and Hu (1989) indicates that this tends to occur at an early stage of the project cycle, such as identification. After that, even at the appraisal stage, little is typically said about the technological aspects of the project. If African countries are to have a countervailing influence on technology choice, therefore,

> They will need to show a much greater degree of concern for technology, at a much earlier stage of the project cycle. Whether, and to what extent, this is happening, is an important topic for research, as is a review of the appropriateness of Chinese aid to Africa.
>
> (James, 2017: 79)

Partly because the Chinese are not given to imparting information about their aid projects in Africa, there is much to be gleaned from such research, which, as noted numerous times in this chapter, has become merely a topic from the past.

Conclusions

After having constituted a major topic in development research in Africa during the 1970s, 1980s and part of the 1990s, appropriate technology has virtually disappeared from the literature during the past twenty years or so. It is not entirely clear what has brought about this drastic decline, but it seems to be related to certain institutional changes and a move towards research in information technology (which of course can also be appropriate, though much of the discussion of it falls into a different analytical framework). Nor can I rule out the possibility that some form of saturation in the received literature has taken place.

The question posted in the title of the chapter is whether this actually matters. It would matter less, for example, if the early research had provided clear answers to the main questions that had been asked of it. And indeed, with respect to the key question of whether this form of technology actually exists across a wide range of manufacturing industries, an affirmative answer has already been widely reported. Such research, however, only provides an intermediate answer to a much larger problem, namely, of ensuring that appropriate technology is widely diffused across a large number of potential users. There are only a few examples of where this has actually occurred in Africa and much research remains to be done on why so little progress has been made on replicating, at a higher level of aggregation, examples that work only on a small (micro) scale.

This type of research, moreover is becoming all the more pressing, as the process of industrialisation envisaged by Lewis turns out to be increasingly more difficult. On the supply side, for example, the population of sub-Saharan Africa is expected to double by the year 2050[19] (at a rate that is much higher than any other region), not to speak of the possible displacement of local firms and workers by low-cost Chinese imports. On the demand side, the prospects are not much brighter, as automation threatens jobs that were once performed labour-intensively. Taken together, these tendencies suggest that appropriate technology research may now have become even more important a topic than it was in the earlier periods cited above.

A similar conclusion seems plausible when one considers the implications of what may be called a 'new international economic order', that is, the emergence in recent decades of a growing Chinese (and Indian) share of R&D, foreign investment, trade and aid resources to sub-Saharan Africa. This shift, I argued, presents a potentially important opportunity for employment creation in the region, in that developing country products and technologies tend to be more appropriate than those from developed countries (though I also emphasised that this is by no means inevitable). Chinese R&D, for example, is sometimes said to be based more on local (Chinese) problems than it used to be and so too, the argument goes, will the solutions to those problems reflect developing-rather than developed-country conditions. There is also some limited evidence that Chinese private investment in Africa is relatively labour-intensive, but here again, the issue warrants far more research attention than it has so far received (as also does the relative factor intensity of Chinese aid projects).

Briefly stated, then, my conclusion is that the near total decline of research on appropriate technology in recent decades is not justified; indeed, there are good reasons to suppose that, in certain key areas, it has become even more urgent than it was in earlier periods.

Notes

1 This paper draws in part on James (2012), but it asks a question not posed in that source. There is a reference to this decline in Huq (2015).
2 Low quality is compensated for by a low price, but only up to a point.
3 There is, however, some literature on the subject. See James (1989) and Stewart (1987); see also below.
4 In 1992 the ILO published a review of the impact of WEP research which covered the topic of appropriate technology in various contexts such as manufacturing and road construction. Impact referred to the influence of such research on academics, policy-makers and library users. See ILO (1992).
5 Note, however, that the framework depicted in Figure 3.1 has been widely criticised. It has been argued, for example, that the points on the isoquant do not represent identical products; that factor prices in developing countries are distorted; that the choice of technique is made in three rather than, as in the figure, just two dimensions; and that those who make the choice have other goals than just employment and efficiency. See Huq (2015) for more details.
6 See Pack (1982).
7 For some such studies, see Stewart et al. (1992).
8 Pack (1982) does note, however, that the labour-intensive alternatives will tend to be associated with higher managerial costs. These would be picked up in a three-dimensional choice of technique framework (James, 1999).
9 Reference must also be made here to the merger of Appropriate Technology International, an American NGO, with Enterprise Works, which was less concerned with research on appropriate technology itself.
10 For a thorough review of WEP research on technology see ILO (1992).

11 The perfectly elastic supply curve of labour is due to the presence of surplus labour in the non-modern part of the economy.
12 Though, in cases of distorted factor prices, this is not necessarily true.
13 This idea was originally advanced by Burenstam-Linder (1961) and it was later developed, most notably, by Stewart (1977).
14 The Mwanza textile mill used in Tanzania in the 1970s is a case in point. See James (1995).
15 For price disparities across industrial sectors see Edwards and Jenkins (2013).
16 In 2016, 'more than half of all active youth' were 'expected to remain unemployed in 2016, representing the highest youth unemployment rate in the region' (ILO, 2016).
17 There are some exceptions, however. See James (1983, 2000) and Ginneken and Baron (1984) for examples.
18 See UN (2015) on the details of this projection.
19 This estimate is due to Kaplinsky (2011).

References

Amsden, A. (1977). The division of labor is limited by the type of market: The case of the Taiwanese machine tool industry, *World Development*, 5,3:217–233.

Bhalla, A.S. (ed.) (1975). *Technology and Employment in Industry*, Geneva: ILO World Employment Programme Research.

Burenstam-Linder, S. (1961). *An Essay on Trade and Transformation*, Stockholm: Almquist and Wiksell.

Carr, M. (ed.) (1985). *The AT Reader: Theory and Practice in Appropriate Technology*, London: Intermediate Technology Publications.

Chandy, L., Hosono, A., Kharas, H. and Linn, J. (eds.) (2013). *Getting Up to Scale: How to Bring Development Solutions to Millions of Poor People*, Washington, DC: Brookings Institution Press.

Edwards, L. and Jenkins, R. (2013). The Impact of Chinese Import Penetration on the South African Manufacturing Sector, Southern Africa Labour and Development Research Unit Working Paper, No. 102, University of Cape Town.

Fransman, M. (1984). Some hypotheses regarding indigenous technological capability and the case of machine production in Hong Kong, in M. Fransman and K. King (eds.) *Technological Capability in the Third World*, New York: St. Martins.

Ginneken, W. van and Baron, C. (eds.) (1984). *Appropriate Products, Employment and Technology*, London: Macmillan.

Huq, M. (2015). Is the choice of technique debate still relevant?, in J. Weiss and M. Tribe (eds.) *Routledge Handbook of Industry and Development*, London: Routledge.

Hyman, E. (1993). Production of edible oils for the masses by the masses: The impact of the RAM Press in Tanzania, *World Development*, 21:429–443.

ILO (1992). *The Impact of World Employment Programme Research on Technology*, Geneva: ILO.

ILO (2016). *World Employment and Social Outlook*, Geneva: ILO.

James, J. (1983). *Consumer Choice in the Third World*, London: Macmillan.

James, J. (1989). *Upgrading Traditional Rural Technologies*, London: Macmillan.

James, J. (1995). *The State, Technology and Industrialisation in Africa*, London: Macmillan.

James, J. (1999). Trait-taking versus trait-making in technical choice: The case of Africa, *Journal of International Development*, 11,6:797–810.

James, J. (2000). *Consumption, Globalization and Development*, London: Macmillan.

James, J. (2016). *The Impact of Mobile Phones on Poverty and Inequality in Developing Countries*, Heidelberg: Springer.

James, J. and Stewart, F. (1981). New products: A discussion of the welfare effects of the introduction of new products in developing countries, *Oxford Economic Papers*, 33,1:81107.

Jéquier, N. and Hu, Y. (1989). *Banking and the Promotion of Technical Development*, London: Macmillan.

Kaplinsky, R. (2011). Schumacher meets Schumpeter: Appropriate technology below the radar, *Research Policy*, 40,2:193–203.

Landes, D. (1965). Japan and Europe: Contrasts in industrialization, in D. Lockwood (ed.) *The State and Economic Enterprise in Japan*, Princeton, NJ: Princeton University Press.

Lewis, W.A. (1954). Economic development with unlimited supplies of cheap labour, *The Manchester School*, 2,2:139–191.

Lewis, W.A. (1979). The dual economy revisited, *The Manchester School*, 47,3:211–229.

Linn, J., Hartmann, A., Karas, H., Kohl, R. and Massler, B. (2010). Scaling Up the Fight against Rural Poverty, Global Economy and Development, Working Paper 43, The Brookings Institution.

Mason, A. (2001). *Population change and economic development: What have we learned from the East Asia experience?*. Paper Presented at the Meetings of the Western Economic Association, San Francisco, July 5–9.

McCarthy, C. (2005). *Productivity Performance in South Africa*, Vienna: UNIDO. Available at www.unido.org/uploads/tx_templavoila/Productivity_performance_in_DCs_South_Afri ca.pdf. Accessed 13 August, 2013.

Pack, H. (1982). Aggregate implications of factor substitution in industrial processes, *Journal of Development Economics*, 11,1:1–38.

Pigato, M. and Tang, W. (2015). *China and Africa: Expanding Economic Ties in an Evolving Global Context*, Investing in Africa Forum, Addis Ababa: The World Bank.

Ranis, G. (1973). Industrial sector labor absorption, *Economic Development and Cultural Change*, 21,3:387–408.

Singer, H. (1970). Dualism revisited: A new approach to the problems of the dual society in developing countries, *The Journal of Development Studies*, 7,1:60–75.

Stewart, F. (1977). *Technology and Underdevelopment*, London: Macmillan.

Stewart, F., Lall, S. and Wangwe, S. (eds.) (1992). *Alternative Development Strategies in Sub-Saharan Africa*, London: Macmillan.

UN (2015). *World Population Prospects: The 2015 Revision*, New York: United Nations.

4   An institutional critique of measures to compare technological capabilities between rich and poor countries

The case of Africa

The early 2000s saw the introduction of indexes which sought to compare countries on a range of technology indicators. In 2006, I argued that such indexes were biased in favour of indicators pertaining to 'developed-country' circumstances (James, 2006). I pointed for example to patents, R&D and 'high-tech' exports, which, while broadly applicable to rich countries, were largely irrelevant to poor, developing countries, such as most of those in sub-Saharan Africa.

Although other models could have been chosen,[1] particular attention was paid to the UNDP's Technology Achievement Index (TAI), which was first introduced in 2001.[2] This index was highlighted because it 'sits so oddly with the well-known emphasis of the UNDP on problems of human development that are disproportionally represented in rural rather than urban areas and from whom one may thus have expected some attention to be paid to the technical achievements in these same areas' (James, 2006). No less important are the two specific goals of the TAI:

> First, to focus on indicators that reflect policy concerns for all countries, *regardless of the level of technological development. Second, to be useful for developing countries.* To accomplish this, the index must be able to *discriminate between countries at the lower end of the range.*
>
> (UNDP, 2001: 46, emphasis added)

The problem, though, is that when it is evaluated on these goals, the TAI performs poorly in discriminating among low-income countries, whose scores on 'high-income' capabilities such as patents, R&D etc. tend to be zero, negligible or unavailable (see below).

This led me to wonder whether a special index is needed, one which would conform much more closely to developing-country needs and circumstances. More specifically, one is bound to ask whether there is a need for a specifically *African* set of capabilities and what these should comprise (Africa being arguably the region most in need of an exercise of this kind). I think there is such a case to be made, and will say more about it in the second part of the chapter. It bears emphasis even at the outset, though, that my discussion revolves around basically two related questions. The first has to do with which countries perform best on selected African capabilities and why. The second question asks how far one can go towards combining these results into something like a collective outcome. I stress that the answers to these questions should be seen very much as tentative: more the starting point of a wider research agenda, on technological capabilities for the sub-Saharan Africa region, than a completed project.

First, however, an important pre-condition has to be met, and it has to do with whether, in the past fifteen to twenty years, African countries have managed to develop their 'rich-country' capabilities such as patents and R&D to the point where they are no longer as irrelevant as they largely were in 2001; for then a single index covering rich and poor countries may indeed be more desirable than it was then. It is to this important preliminary issue that the chapter first addresses itself.

Growth of 'high-income' capabilities in sub-Saharan Africa

It will be argued in this section that with the exception of Internet users, whose numbers have grown rapidly since their near negligible status in 2001, there has been very little growth in 'high-income' capabilities in Africa during the last fifteen to twenty years (see Tables 4.1 to 4.3 which contain the same selected

Table 4.1 Patents per country, selected sample

Country	1985	2014
Benin		
Botswana		5
Cameroon		
Central African Republic		
D.R. Congo		
Ethiopia		
Ghana		
Kenya	98	75
Malawi	35	
Mali		
Mauritius	5	
Mozambique		
Namibia		
Senegal		
Sierra Leone		
South Africa	5,849	6,750
Uganda	26	3
Sub-Saharan Africa		
OECD	n/a	516,028 (2014)

Source: World Bank Indicators, 2015

Note: Blank entries denote that no data were available at the time of the latest update.

Table 4.2 R&D as per cent GDP, selected sample, 1996–2013

Country	1996	2013
Benin		
Botswana		
Cameroon		
Central African Republic		
D.R. Congo		
Ethiopia		6
Ghana		
Kenya		
Malawi		
Mali		
Mauritius		
Mozambique		
Namibia		
Senegal		
Sierra Leone		
South Africa		
Uganda		
Sub-Saharan Africa		
OECD	n/a	2.42 (2013)

Source: World Bank Indicators, 2015

Note: Blanks denote that no data were available at the time of the latest update.

Table 4.3 High-tech exports (as per cent manufactured exports) selected sample, 1989–2015

Country	1989	2015
Benin		1
Botswana		1
Cameroon		4
Central African Republic		
D.R. Congo		
Ethiopia		4
Ghana		
Kenya		
Malawi		2
Mali		
Mauritius		0
Mozambique		12
Namibia		
Senegal		4
Sierra Leone		0
South Africa		6
Uganda		2
Sub-Saharan Africa		
OECD	n/a	17 (2015)

Source: World Bank Indicators, 2016

Note: A blank denotes that no data were available at the time of the latest update.

country samples). To this extent, there is no more justification for using the TAI in this region than there was in 2001.

With certain partial exceptions (such as South Africa in patents and Mozambique in high-tech exports), the entries in the right-hand column of Tables 4.1 to 4.3 lie well below the average for a sample of developed countries (OECD) shown in the last row of the tables. Note that, as far as the second column is concerned, the problem of data unavailability is especially pronounced in regard to patents and R&D (a function itself no doubt of the negligible amounts of the capabilities in question). In all three cases, however, the overall figure for sub-Saharan Africa is denoted as being unavailable. All in all, my conclusion is that African scores on patents, R&D and high-tech exports are still basically as low or unavailable as to make current comparisons with developed countries untenable. An alternative framework is needed, one which is based, as noted above, on basic African skills that are mediated largely through the school system. This framework will be discussed in a later section after I have dealt with the exceptional case of Internet skills.

Internet use as a basic skill

Basic Internet skills in Africa, however, have not lagged nearly as far behind the OECD countries as the other three capabilities described above. On average, for example, there are data for the former region as a whole and they show that Internet use is equal to 22.4 per cent (see Table 4.4). This amount is a far cry, I aver, from the unavailable or negligible estimates for sub-Saharan Africa shown in Tables 4.1 to 4.3, which refer to the three high-income capabilities discussed above. In fact, I am inclined to include digital competence as one of the four basic learning skills to be described as part of an 'African' set of basic capabilities (the subject of the following section). After all, it is becoming increasingly important for candidates in the African manufacturing job market to possess some basic level of digital skills. Mobile phones do not represent a viable alternative in this regard, because they only require the bare minimum of such skills.

Beginning, uniformly, from a position of zero in 1990, sub-Saharan African countries differed only very slightly from OECD countries in that year, a time when the Internet scarcely existed. For the moment, all I wish to say about the data in the right-hand column of Table 4.4 is that they are not at all rife with unavailabilities, zeros and negligible amounts, as were the comparable amounts in Tables 4.1 to 4.3. This discriminant feature of Table 4.4 is salient because of the comparisons it allows between African countries, between sub-Saharan Africa as a whole and other regions such as the developed countries (represented by the OECD). See below for a comparison of how African countries in the sample given by Table 4.4, have performed on the basis of digital skills, with particular reference to the exceptional cases.

Table 4.4 Internet users per 100 persons, 1990 and 2015, sub-Saharan Africa

Country	1990	2015
Angola	0	12.4
Benin	0	6.8
Botswana	0	27.5
Burkina Faso	0	11.4
Burundi	0	4.9
Cameroon	0	20.7
Central African Republic	0	4.6
Chad	0	2.7
DR Congo	0	3.8
Rep. Congo	0	7.6
Cote d'Ivoire	0	21.0
Djibouti	0	11.9
Eritrea	0	1.1
Ethiopia	0	11.6
Gabon	0	23.5
The Gambia	0	17.1
Ghana	0	23.5
Guinea	0	4.7
Guinea-Bissau	0	3.5
Kenya	0	45.6
Lesotho	0	16.1
Madagascar	0	4.2
Malawi	0	9.3
Mali	0	10.3
Mauritania	0	15.2
Mauritius	0	50.1
Mozambique	0	9.0
Namibia	0	22.3
Niger	0	2.2
Nigeria	0	47.4
Rwanda	0	18.0
Senegal	0	21.7
Seychelles	0	58.1
Sierra Leone	0	2.5
South Africa	0	51.9
South Sudan		17.9
Sudan	0	26.6
Swaziland	0	30.4
Tanzania	0	5.4
Togo	0	7.1
Uganda	0	19.2
Yemen	0	25.1
Zambia	0	21.0
Zimbabwe	0	16.4
Sub-Saharan Africa	0	22.4
OECD	0.2	77.2

Source: World Bank Indicators, 2016

Implications for an 'African' set of capabilities

The empirical evidence in the previous section leads me to conclude that there is a need for an alternative 'African' set of skill-based capabilities. Such an index will hopefully allow meaningful discrimination between countries in the region, a task that was set but not fulfilled in the original TAI concept (as already noted). My choice in this regard is a set of basic learning skills which are more or less successfully acquired through the school system (supplemented in certain cases by other institutions as well). It is not that there are no skills indicators in the original TAI or its updated version in 2009.[3] Rather, the problem is that what is included there is conceptually flawed, and ignores new data sets that have emerged in recent years. Certainly, among the low-income majority in typically unequal African societies, it is basic skills that are crucial in determining whether or not individuals are able to escape from pervasive poverty. It is in this sense that the discussion about 'African' capabilities is to be understood, as distinct from being about the minority who are reliant on modern methods and skills.

As far as my basic skills categories are concerned, the intention is to describe and discuss them and to discern which countries perform well on each one (as well as to compare how well they do on these categories as a whole). I begin with digital skills not because they are necessarily the most important, but rather because they admit explanation more easily than the other capabilities (numeracy, literacy and vocational training at secondary school).

Consider, in particular, the list of Internet users by country per 100 persons in Table 4.4. A cursory glance at this table suggests that the best-performing countries, such as the Seychelles, are also among the wealthiest. Table 4.5 shows this pattern more clearly by comparing the top-five best performers on Internet use with their per capita incomes. Apart from Kenya, the other countries have per capita incomes above the average for the region (and in some cases, significantly so).

Other, more rigorous regression studies in developing countries have confirmed the link between income per capita and Internet use (e.g., Chinn and

Table 4.5 Top-five Internet users and per capita incomes

Country	Internet use per 100 (2015)	Per capita income
Seychelles	58.1	15,476.0
South Africa	51.9	5,724.0
Mauritius	50.1	9,252.1
Nigeria	47.4	2,640.3
Kenya	45.6	1,376.7
Sub-Saharan Africa	22.4	1,1588.5

Source: First column Table 4.4 and second column World Bank Indicators

Note: current US$ 2015

Fairlie, 2004). This should not come as a particular surprise, however, since there are plausible explanatory factors on both the supply and demand sides. As regards the former, there are often heavy costs associated, for example, with satellites and other infrastructure, while with respect to the latter (the demand side) low incomes hinder the adoption of even 3G Internet services.

The exceptional performance of Kenya at a slightly lower than average per capita income, however, needs to be explained in other terms. These, I would suggest, have to do partly with a favourable technology culture, including state policy towards new technology in general and fibre-optic cables more specifically. Nairobi, for example, has a reputation as being one of the few 'technological hubs' in sub-Saharan Africa, exhibiting, as it does, relatively favourable attitudes towards technological innovations in general and information technology in particular[4] (consider in this respect the success of the well known mobile money scheme known as M-Pesa, which has brought mobile banking to tens of millions of Kenyans, even those in remote locations[5]).

Income, however, is much less effective as an explanatory variable when it comes to the other three capabilities, namely, literacy, numeracy and percentage of secondary school students with vocational training. Note, to begin with, that the data on literacy and numeracy were collected outside the usual international collection process by a regional institution, Pasec, focusing on West Africa.[6] As educational achievements they constitute a substantial improvement over the imperfect measure of school enrolment used in the original and 2009 versions of the TAI.

The essential point is that the one concept is really an input to the process by which literacy and numeracy (as outputs) are actually generated. Sen has made much of this distinction with his notion of 'functionings' (Sen, 1985). He sees products in general as inputs in a process which culminates in a set of achievements or 'functionings'. The same product can yield very different outcomes to different users.[7] In the case of medicinal drugs, for example, one patient may experience the expected therapeutic gains while another, who, say, uses medicine in the incorrect way, experiences mostly harmful side-effects (or 'negative' functionings). Or again, in the educational context, enrolment may translate into different functionings among students in the same school, or even the same class, depending, among other things, on whether he or she gets help with homework from parents or siblings.[8]

For far too long, however, those concerned with education policy in Africa have wrongly focused on the number of pupils in schools rather than, or in addition to, what they actually achieve while there.[9] And when the situation is viewed empirically from this alternative, more modern point of view, the results are often disturbing. For instance, research by the Brookings Centre for Universal Education has estimated that '61 million African children will reach adolescence lacking even the most basic literacy and numeracy skills. Failure to tackle the learning deficit will deprive a whole generation of opportunities to develop their potential and escape poverty' (Watkins, 2013: 1).

Data at the more disaggregated level of countries show just how severe the failure to meet minimum mastery levels in literacy and numeracy can be. Thus,

> Over one third of pupils covered in the survey – 23 million children – fall below the minimum learning threshold. Because this figure is an average, it obscures the depth of the learning deficit in many countries. More than half of students in Grades 4 and 5 in countries such as Ethiopia, Nigeria and Zambia are below the minimum learning bar. In total, there are seven countries in which 40 percent or more of children are in this position. As a middle-income country, South Africa stands out. One third of children fall below the learning threshold, reflecting the large number of failing schools in areas servicing predominantly low-income black and mixed-race children.
>
> (Watkins, 2013: 7)[10]

But whereas the examples in the citation refer to failed instances of basic learning achievements, it is important to look also at the cases that can be deemed successful. Thus, in Table 4.6, I have listed the results for a sample of SSA African countries (recall that because the entries refer to the percentage who

Table 4.6 Learning levels (numeracy and literacy), selected sample of African countries

Country	Literacy (per cent)	Numeracy (per cent)	Composite (per cent not meeting basic learning level)
Benin	44.8	38.5	41.7
Burkina Faso	31.4	24.9	28.2
Burundi	16.6	15.5	16.1
Cameroon	9.0	10.2	9.6
Chad	45.0	34.9	40.0
Comoros	37.5	30.8	34.2
Congo	37.9	31.9	34.9
Ethiopia	54.2	56.3	55.3
Gabon	6.2	10.9	8.6
Ghana	21.1	43.1	32.1
Ivory Coast	33.6	48.3	41.0
Lesotho	21.2	41.8	31.5
Madagascar	23.6	6.5	15.1
Malawi	36.6	59.9	48.3
Nigeria	65.7	51.0	58.3
Senegal	24.0	19.2	21.6

Country coverage: (PASEC) Chad, Benin, Comoros, Madagascar, Gabon, Burkina Faso, Congo, Senegal, Burundi, Côte d'Ivoire and Comoros; (National examinations) Ghana, Ethiopia and Nigeria.

Sources: PASEC Watkins (2013)

Table 4.7 Per capita income of top-five ranked countries
on literacy and numeracy, 2015

Country	Per capita income
Cameroon	1,217.3
Kenya	1,376.7
Madagascar	401.8
Mauritius	9,252.1
Gabon	8,266.4
Swaziland	3,200.1
Tanzania	879.0
Sub-Saharan Africa	1,588.5

Source: World Bank indicators

fail to receive minimum mastery levels, lower numbers are superior as learning functionings to higher ones). There are also results for a comparable sample from East and South African countries, which I list for numeracy and literacy as follows: Botswana, 22.4, 10.6; Chad 34.9, 45.0; Kenya, 11.2, 8.0; Mauritius, 11.2, 11.1; Mozambique, 32.8, 21.5; Namibia, 47.7, 13.6; Seychelles, 17.8, 11.8; South Africa, 40.2, 27.2; Swaziland, 8.6, 1.4; Tanzania, 26.6, 18.5; Uganda, 38.8, 20.4; Zambia, 67.3, 44.1; Zimbabwe, 26.6, 18.5 (Sacmeq, III).

Combining the results from West and Southern African countries leads to a group of top-five, as shown in Table 4.7 (together with their per capita incomes).

As a general first impression, income does not seem to play a dominant role in explaining the observed pattern of successful cases. In fact, of the seven countries listed as being in the top-five on either of the achievements, Table 4.7 shows that only three have incomes higher than the sub-Saharan African average.

In early econometric research, Michaelowa (2001) pointed to some of the non-income factors that appeared to favourably influence learning achievements in five West African countries. She pointed, for example, to 'teachers' knowledge of the local language', the help with homework that pupils may receive from parents and (certain) siblings and the motivation of teachers (as reflected for example in the extent of absenteeism).

Another general observation is that in four of the cases shown in Table 4.6, the same country appears in the top five of both rankings. This suggests that similar forces may be at work across achievements, but there is regrettably little *current* research on the determinants of educational achievements in Africa, or of the source of any joint causation. These gaps reflect a rather glaring weakness in the literature, one to which I return in the concluding section.

It is nonetheless worth citing the conclusions from Michaelowa's early research because it bears on findings cited elsewhere in this chapter. She finds, for example, that:

> It becomes clear that the relationship between primary education expenditure and educational outcomes is not straightforward. Apparently, financial

resources are much more efficiently used in some countries than in others. At the higher end of educational outcomes, Cameroon clearly predominates over *Côte d'Ivoire*, and at the lower end, Madagascar and Burkina Faso clearly rank above Senegal in terms of efficiency. These efficiency positions depend on various political, economic, social, geographic and cultural factors.

(Michaelowa, 2001: 170, emphasis added)

This conclusion bears first on and provides evidence for Sen's notion noted above, that the relationship between the spending on and the output of education cannot be taken as rigid, and depends as he suggests on factors drawn from a wide range of disciplines. Income need not play a decisive role. In fact, it will soon become apparent – after discussing the vocational training in secondary education – that only Kenya and Cameroon perform exceptionally well in three out of the four capabilities in question.[11] And if one assumes that what matters is mere ranking in the top group, and not the size of the difference between entries, these two countries can claim in a rough way to be the best performers overall (I return though to an alternative method below). What stands out about Kenya and Cameroon, moreover, is that they both have per capita incomes below the average for sub-Saharan Africa (see Table 4.7). A major lesson from these two countries is thus that relatively high levels of basic skills attainment are possible even at relatively low per capita incomes.[12] Finally, one needs to consider the possibility that Cameroon's stellar recent experience has an affinity with and is indeed partly the outcome of its earlier performance as described by Michaelowa (2001). Traditions often run deep in education, and the possibility I have described may be another example of this tendency towards historical continuity.

Percentage of secondary school students receiving vocational training

The last, but by no means least, of the basic skills under consideration here is the percentage of secondary school students who receive vocational training (see Table 4.8). One of the reasons why it is important is that according to Lall (1992: 117) secondary education is 'The most critical input for industrial development'. What I think he meant by this is that it is the stage where skills are learnt that have a direct bearing on industrial development, such as carpentry or metal-working. Another reason is that the data contained in Table 4.8 are mostly positive and allow even poor African countries to be differentiated from one another. Angola, Cameroon and the Democratic Republic of the Congo receive the highest scores for reasons that are not at all readily apparent. (There is unfortunately very little comparative literature on the subject.)

Note, though, that this is the third out of four possibilities in which Cameroon has appeared in the top five places and it is notable that this country has one of the lowest levels of per capita income among the leading group marked in the table. What also perhaps bears some explanatory emphasis here is that

Cameroon has achieved some degree of industrial success in the past, and this may have called forth the need for vocational skills in the sector, but here again, there is an acute shortage of available research to draw on.

Note, finally, that for some unknown reason, Kenya is not included in Table 4.8. But, if and when it is, and it performs relatively well, this country would become

Table 4.8 Percentage of secondary students with vocational training, Africa (available years)

Country	Year	Number with vocational training at secondary school	Population	Col. 3 ÷ by col. 4 in per cent	
Angola	2011	400,265	21,942,296	1.8	1)
Benin	2011	24,626	9,779,391	0.25	
Burkina Faso	2013	29,730	17,084,554	0.17	
Burundi	2013	26,149	10,465,959	0.25	
Cameroon	2013	383,539	22,211,166	1.7	2)
CAR	2012	3,850	4,619,500	0.08	
Chad	2012	6,855	12,715,465	0.05	
Democratic Republic of Congo	2013	745,343	72,522,861	1.0	3)
Rep. Congo	2012	34,336	4,286,188	0.8	5)
Côte d'Ivoire	2013	29,595	21,622,490	0.14	
Djibouti	2013	2,338	864,554	0.27	
Eritrea	2013	2,470	5,110,444	0.05	
Ethiopia	2012	201,142	92,191,211	0.2	
Ghana	2013	61,496	26,164,432	0.2	
Lesotho	2012	6,691	2,057,331	0.33	
Liberia	2011	17,565	4,079,574	0.4	
Madagascar	2013	50,724	22,924,557	0.22	
Malawi	2012	0	15,700,436	0	
Mali	2013	109,899	16,592,097	0.66	
Mauritania	2013	2,139	3,872,684	0.06	
Mauritius	2012	11,446	1,258,653	0.9	4)
Mozambique	2013	35,397	26,467,180	0.13	
Niger	2013	37,250	18,358,863	0.2	
Rwanda	2013	80,458	11,078,095	0.7	
Senegal	2011	37,516	13,357,003	0.28	
Seychelles	2013	226	89,900	0.25	
South Africa	2013	359,191	53,157,490	0.68	
Sudan	2012	25,167	37,712,420	0.06	
Swaziland	2011	0	1,212,458	0	
Tanzania	2013	248,239	50,213,457	0.5	
Togo	2011	27,573	6,566,179	0.4	

Source: World Bank (2015), Indicators. Last column, author calculations

the sole leader according to the samples I have used, and the method I have thus far employed. As noted below, however, an alternative method needs to be reported, one that is based on a very different premise regarding how countries should be compared. It is to this that I next turn.

An alternative method

According to the method used so far, all the countries in the top five receive positive but equal weights as opposed to the remaining countries which are assigned zero. This method favours breadth in superior performance (for example, Kenya and Cameroon both perform exceptionally well according to three out of the four indicators). In the alternative method, countries outside the top group continue to receive zero scores. What is different is that a distinction is made between countries in this (top) group. In particular, the leading country (arbitrarily) receives five points, the second four points and so on. The results are shown in Tables 4.9 and 4.10, which, taken together, indicate that Swaziland comes first, closely followed by Cameroon, with a group of countries in joint third place (each of which performed best on just one set of capabilities).

Kenya now drops entirely out of the picture because of its relatively weak performance on the capabilities where it does appear in the top five. This case illustrates well the difference between the two methods that are being compared.

Table 4.9 Weighting of countries in top five, all capabilities[1]

Numeracy		Literacy		Vocational		Internet use	
Swaziland	5	Madagascar	5	Angola	5	Seychelles	5
Tanzania	4	Swaziland	4	Cameroon	4	South Africa	4
Gabon	3	Cameroon	3	D.R. Congo	3	Mauritius	3
Kenya	2	Gabon	2	Mauritius	2	Nigeria	2
Cameroon	1	Kenya	1	Rep. Congo	1	Kenya	1

1) Five is the maximum, and one is the minimum score.

Table 4.10 Overall ranking by alternative method

Country	Score
Swaziland	9
Cameroon	8
Madagascar	5
Angola	5
Seychelles	5
Mauritius	5

Source: based on Table 4.9.

In contrast to the first method, which favours breadth of good performance, the second favours narrowness of superior achievements. Swaziland, for example, comes top on the basis of exceptional scores on only two capabilities. Being in both numeracy and literacy, moreover, these scores may be related in part to the same causes. In particular, primary school in that country is fully financed by the government. Students receive free textbooks, stationery and meals, each of which was found to be significant in explaining numeracy and literacy by Michaelowa (2001) in the study referred to above.

Ultimately, of course, countries need both breadth and maximum scores on each capability to derive the highest possible number of points, which is equal to twenty. That this number is nowhere near to being reached in Table 4.9 is due, logically, to two possibilities. One is that there are countries in all four areas, but which simply do not perform at the required level. The other is that certain countries are missing from one or more data categories (such as literacy or numeracy). After all, these particular data sets comprise just twenty-eight out of all the countries in the sub-Saharan Africa region (although Internet use, as shown in Table 4.4, contains a distinctly wider variety of countries).

Ultimately, though, the purpose of this exercise was not so much to explain why countries fail to attain the maximum score, as it was *to draw attention to more limited successes*. I am referring here for example to Cameroon and Kenya's breadth of strong performance on three out of the four dimensions I have chosen, and to Swaziland's outstanding record in the combination of literacy and numeracy scores. It is also salient to note that in all these cases, country per capita income was below the average for the region as a whole, suggesting that important non-income explanatory forces were at play (some of the suggested ones have been mentioned above). After all, these forces are often of a socio-political character and are heavily subject to government policy. On the other hand, Internet use is a relatively expensive proposition, and the demand for service appears to be correlated with the per capita incomes of countries.

Conclusions

Basing my analysis on the failure of most indicators of technological capabilities (such as the TAI) to meaningfully distinguish between poor developing countries, I have suggested that alternative indexes be considered, ones which are based on the specific characteristics of poor, rather than rich countries, especially in Africa.

I showed that the contribution of African countries to 'developed-country' capabilities such as patents, R&D and high-tech exports has remained largely negligible or unavailable for the fifteen-odd years since the TAI was first introduced. An alternative group of capabilities was therefore proposed instead. These involve basic learning skills – mediated mainly through the school system – which are vital for the poor majority in the typical African country to escape from poverty and inequality. Four such capabilities in particular are identified:

namely, numeracy and literacy, vocational skills in secondary education and Internet use (which is partly, of course, also a developed-country capability).[13]

Using some relatively recent data sources, I find that no one country achieves a leading position on all four capabilities, even those that are relatively affluent. This is partly due to the fact that it is a very demanding requirement and partly a question of data unavailability. Countries which do well on certain indicators may simply be unavailable when it comes to other dimensions. That is to say, there is an incomplete overlap of countries between data sets. Nonetheless, two countries, Kenya and Cameroon, find prominent positions on three out of the four possibilities (Swaziland does best, though, when a method is used that discriminates between countries in the top tier of a category).

It is notable that all three of the cases I have just mentioned exhibit per capita incomes below the average for Africa as a whole. What is suggested thereby is that there may be important non-income forces at work in determining how well countries perform on basic-skill capabilities.[14] Some such reasons, including parental help with homework and teachers' motivations, were identified in early research on West Africa. Unfortunately, there is little light that can currently be shed on the question, and it is to be hoped that future research will help to fill what is a rather glaring gap in the literature and to address the more general question I have introduced in this chapter: namely, of measuring technological capabilities in the specifically African context of poverty and inequality.

Some of the gaps will involve case studies of countries that are already known to be successful in some areas, such as Cameroon in literacy and numeracy. Other studies, however, will involve cross-country econometric work to determine (say) the explanations for the rankings according to the percentage of students that receive vocational training in secondary schools.[15] In this particular exercise, political variables are likely to be just as important as economic ones.

Notes

1 For a list of alternatives, see Archibugi and Coco (2005).
2 In the Human Development Report of 2001.
3 The author of the updated exercise was Nasir (2009).
4 See Wangalwa (2014) on Kenya's policy towards the Internet.
5 See James (2016) for a summary of the M-Pesa experience.
6 For a useful description of the process, see Watkins (2013).
7 Sen himself uses the example of food, which has a variety of nutritional and sociological characteristics (Sen, 1985).
8 For a discussion of other relevant variables, see Michaelowa (2001). This article is referred to at several different points in the chapter.
9 This point is emphasised by Watkins (2013).
10 This country would thus probably be described as an outlier when educational achievements are plotted against per capita incomes.
11 No country performs exceptionally well on all four cases.
12 Conversely, South Arica shows that poor performance is possible even at relatively high incomes.
13 Although in developed countries, skills include much more complex operations.

14 There is an analogy here with the Human Development Index of the UNDP. Some countries, such as Sri Lanka, perform much better in terms of human development indicators than their income would allow. On the other hand, there are countries such as Brazil which do much worse than is expected in terms of their incomes.
15 In this, researchers might find it useful to draw on the methodology used by Michaelowa (2001). There is also more recent work by this author and her colleagues which may be useful, namely, Fehrler et al. (2009).

References

Archibugi, D. and Coco, A. (2005). Measuring technological capabilities at the country level: A survey and a menu for choice, *Research Policy*, 34,2:175–194. doi.org/10.1016/j. tespol.2004.12.002.

Chinn, M. and Fairlie, R. (2004). The Determinants of the Global Digital Divide: A Cross-Country Analysis of Computer and Internet Penetration, National Bureau of Economic Research, Working Paper No. 10686, Boston, MA: USA.

Fehrler, S., Michaelowa, K. and Wechtler, A. (2009). The effectiveness of inputs in primary education: Insights from recent student surveys for Sub-Saharan Africa, *Journal of Development Studies*, 45,9:1545–1578. doi.org/10.1080/00220380802663625.

James, J. (2006). An institutional critique of recent attempts to measure technological capabilities across countries, *Journal of Economic Issues*, 40,3:743–766. doi.org/10.1080/00213 624.2006.11506943.

James, J. (2016). *The Impact of Mobile Phones on Poverty and Inequality in Developing Countries*, Heidelberg: Springer.

Lall, S. (1992). Structural problems of African industry, in F. Stewart, S. Lall and S. Wangwe (eds.) *Alternative Development Strategies in Sub Saharan Africa*, London: Macmillan.

Michaelowa, K. (2001). Primary education quality in Francophone Sub-Saharan Africa: Determinants of learning achievement and efficiency considerations, *World Development*, 29,10:1699–1716. www.sciencedirect.com/science/piis0305750x01000614.

Nasir, A. (2009). Technology achievement index 2009: Ranking and comparative study of nations, *Scientometrics*, 87,1:41–62. Available at www.link.springer.com/article/10.1007/ 511192-010-0285-6.

Sen, A. (1985). *Commodities and Capabilities*, Amsterdam: North Holland.

UNDP (2001). *Human Development Report*, Oxford University Press: New York.

Wangalwa, E. (2014). Kenya Leads Africa's Internet Access and Connectivity, *CNBC Africa*. Available at www.cnbcafrica.com/news/east-africa/2014/09/09/kenya-leads-internet/. Accessed January 24, 2015.

Watkins, K. (2013). *Too Little Access, Not Enough Learning: Africa's Twin Deficit in Education*, Washington, DC: Brookings Institution Press. January 16. Available at www.brookings. edu/opinions/too-little-access-not-enough-learning-africas-twin-deficit-in-education/). Accessed 15 July, 2015.

World Bank (2015). *Manufacturing FDI in Sub-Saharan Africa*, Washington, DC: World Bank.

Part II
Digital technologies

5 Internet use, institutions and well-being

Evidence from Africa

For a very long time, welfare (or quality of life) in economics has been measured by the (nature and) amount of goods consumed. Implicitly, this practice has assumed and continues to assume that welfare is conferred to consumers at the point where goods are purchased and utility is gained. More recent theories, however, such as Sen's (1985) functionings approach, suggest that what also matters to well-being is the use to which goods and technologies are put after purchase. It is one thing, for example, to adopt a product but its ultimate value depends, inter alia, on how and how intensively it is used after purchase. Such information however is difficult to obtain and requires survey methods to collect. In the case of information and communication technology (ICT) in developing countries, surveys of this kind are few and far between and those that do exist, relate, as far as I am aware, entirely to mobile phones.[1]

Over the years, though, rigorously collected and detailed Internet data for eleven African countries have become available, and they shed quite some light on the issues at hand.[2]

The task below is accordingly to present such data and where possible to explain them. Among the questions that are asked in this regard are: do the patterns of use favour one particular group of African countries over others? Which use mechanisms are most important across the sample countries? How do these results compare with those of a developed country, such as the United States? Which variables (such as income and education) correlate most closely with observed Internet use? Before setting out to answer these and other questions, however, I begin with a brief description of the basic model and show how it compares with traditional theory. This is followed by a short description of the survey itself. The analysis proper then takes place by averaging across rows and columns of the basic data that are presented as an extended table. Stepping outside this framework, I examine next the effect of Internet use on social capital in different countries and calculate a macro adjustment to the data. With an eye to possible policy implications, the final section investigates the major constraints on Internet use in the sample countries.

The differences between the traditional economic approach and the use-based model can be summarised as in Figure 5.1. In the former, access to the Internet takes place exclusively by means of ownership and utility is somehow

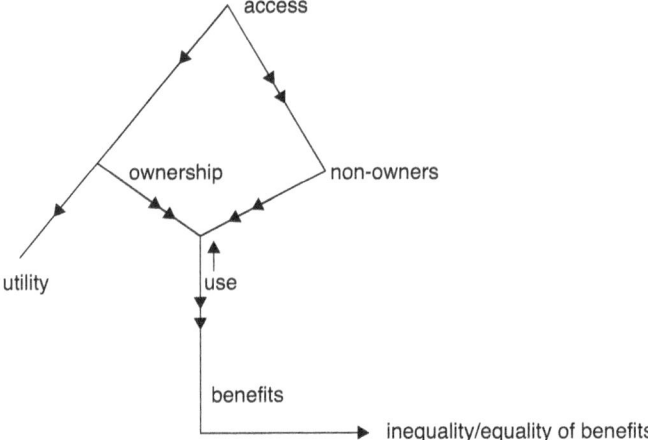

Figure 5.1 Alternative models of consumption and well-being

Note ▶ = traditional model. ▶▶ = use-based model.

Source: James (2014)

Table 5.1 Nonowning users of Internet facilities (illustrative countries)

Location	Uganda per cent of total locations where Internet used	Tanzania per cent of total locations where Internet used
Home	5.6	15.5
Work	13.2	14.1
Place of education	12.3	7.7
Another person's home	13.0	7.6
Community Internet access	10.3	9.9
Commercial Internet access	17.8	19.9
Any place via mobile phone	19.5	23.7
Any place via another mobile access device	8.2	1.5

Source: Research ICT Africa (2011).

derived therefrom (with no attention paid to what is actually done with it). The latter, by contrast, examines both owning and nonowning users (such as visitors to a cybercafé) and is directly concerned with the ways in which Internet use provides actual benefits to both groups (and the inequality or equality of the distribution of such benefits).[3]

Table 5.1 shows the variety of ways in which the Internet is accessed in the two illustrative countries, Uganda and Tanzania. From the point of view of non-owning users, what are jointly most important are community and commercial access points (with the latter taking the form generally of cybercafés). In both countries, these forms of access together account for nearly 30 per cent of all locations. Note that this is only a minimum estimated amount of nonowning users, for there may be others who share or rent this technology at other people's places and at their places of work and education. Note further that sharing is known to take place in mobile phone use, thus further expanding the pool of possible nonowning users.[4]

The benefits of use

Even if nonowners are included in a sample, as noted previously, there remains the vital question of how users actually benefit from the Internet. Traditional economic theory is not very helpful in this respect because it assumes that welfare or well-being occurs at the point where the technology is purchased (from where a certain measure of utility is derived). As Sen (1985) and others have persuasively argued, however,

> What matters for well-being is not just the characteristics of commodities consumed, as in the utility approach, but what use the consumer can and does make of commodities. For example, a book is of little value to an illiterate person (except perhaps as cooking fuel or as a status symbol). . . . To make any sense of the concept of human well-being in general, and poverty in particular, we need to think beyond the availability of commodities and consider their use: to address what a person does (or can do) with the commodities . . . that they come to possess or control.
>
> (Todaro and Smith, 2011: 16)

The purpose of this chapter, accordingly, is to move in the direction thus proposed, by studying the patterns of use in eleven African countries as described in a detailed survey, from 2011. First, however, I briefly discuss the method used in the survey, which was conducted by 'Research ICT Africa', a South African research institution.

Survey methodology

There are several levels on which the method adopted by the survey can be described as impressive. In the first place, there is a diverse range of eleven African countries, running from the very poor to those among the richest on the continent (see Table 5.2). The selected countries are also dispersed by region including East, West and Southern Africa. At the level of each country, moreover, the survey is based on relatively large numbers of respondents, as shown in Table 5.3.

In a related fashion, the data are gathered with the use of national statistical methods that help to ensure nationally representative information (e.g., national census sample frames and enumerator areas). Note, finally, that the survey covers

Table 5.2 Correlating Internet use with selected variables

Country	Use source	Per capita income ($)	Tertiary enrolment	Fixed broadband Internet access tariff ($ a month)
Namibia	65.6	7,500	5.9	47.2
Uganda	61.7	1,300	3.0	194.4
Ghana	60.7	3,100	3.3	44.4
Nigeria	58.8	2,600	4.3	104.0
Kenya	52.9	1,800	3.0	39.8
Botswana	52.7	16,200	4.6	62.2
Tanzania	51.4	1,500	0.7	63.6
Cameroon	49.0	2,300	4.9	88.6
South Africa	45.5	11,100	15.2	26.9
Rwanda	43.5	1,400	1.7	88.0
Ethiopia	34.5	1,100	1.6	496.5
Correlation with use	–	.09	–.006	–.54

Source: CIA World Factbook: World Bank, Indicators; The Little Data Book on Information and Communication Technology: Research ICT Africa, own calculations

Table 5.3 Number of households interviewed by country

Botswana	900
Cameroon	1,200
Ethiopia	1,600
Ghana	1,200
Kenya	1,200
Namibia	900
Rwanda	1,200
South Africa	1,600
Tanzania	1,200
Uganda	1,200

Source: Research ICT Africa (2011)c

an impressively large number of ways in which the Internet is used to convey benefits to users in the countries concerned.

The basic data

The bulk of the data collected in the country surveys can be presented in Table 5.4, with the columns representing the eleven countries and the rows showing the extent to which the Internet is used in a range of different mechanisms. The cell in the first row and column, for example, indicates that 50.7 per cent of Internet users in Uganda employ this technology for getting information

Table 5.4 The basic data: per cent who sometimes engage in mechanism

Mechanism	Country										
	Uganda	Kenya	Tanzania	Rwanda	Ethiopia	Ghana	Cameroon	Nigeria	Namibia	South Africa	Botswana
Getting information about goods or services	50.7	53.4	65.1	35.5	35.1	53.9	63	62.9	70.3	49.8	50.7
Getting information related to health and health services	64.1	69.8	57.8	27.8	27.8	78.8	53.9	71.9	66.9	39.5	65.2
Getting information from government organisations	52.9	56.8	43.1	72.5	72.5	53.4	45.4	60.3	56.3	31.5	50.5
Interacting with government organisation	39.5	28.7	33.8	26.2	34.2	35.8	28.2	47.9	48.5	22.4	35.3
Sending or receiving e-mail	88	89.1	89.3	90.8	64.7	96.2	89.8	83.5	77	68.3	77.1
Telephoning over the Internet (VoIP)	54	26.8	14.6	10.4	13.3	45.7	0.4	32.2	47	24.1	28
Posting information or instant messaging	68.7	80.8	68.8	80.4	52.9	84.5	67.7	75.4	50.4	61.7	71.3
Purchasing or ordering goods and services	39	17	18	19.9	0.9	18	16.4	39.6	39.3	25.4	21
Internet banking	11.1	13.4	23.7	10.4	0	19.9	10.8	24.7	42.4	32.4	22.9
Education or learning activities (formal)	74.9	61.5	50.2	26.9	35.5	87.3	47.9	69.8	69.3	52	56.5
Playing or downloading video games or computer games	72.4	71.8	63.9	28.8	34.2	83.4	50.9	65	79	0.3	56.3
Downloading movie images, music, watching TV or video or listening to radio or music	81.4	76.2	80.6	39.1	39.3	86.5	50.7	63.4	83.7	59.7	66.6
Downloading software	55.4	53.3	51	34.4	28.9	74.5	27.5	44.3	61.1	40.3	43.1
Reading or downloading online newspapers, magazines or electronic books	66.8	64.8	53.8	57.4	75.3	45	39.5	60.7	84.6	43.7	69.8
Participating in distance learning for an academic degree or job training	57.2	24.7	29	18.4	6.9	32.3	4.4	30.8	54.3	28.7	26.3
Getting information for school-or university-related work / researching a topic	73.7	63.4	50.1	60.3	40.8	66.6	53.3	73.2	76.9	56.5	60.2
Looking for free education content, such as free courses	71	64.9	40.9	32.2	22.7	37.5	58.4	54.6	71.6	43.3	53.3
Collaborating online on documents (e.g., Google and OCS)	70.2	51.5	50	42.8	41	55.6	59.2	64.3	63.6	50.2	46.6
Social networking or video-sharing websites	69.8	82.6	58.9	81.1	37.5	81.1	72.4	76	81.1	70.6	72.5
Finding or checking a fact	72.8	77	80.2	75.6	64.1	77.4	70.8	75	88	60.9	80

Note: VoIP = Voice over Internet Protocol; OCS = Office Communications Server

Source: Research ICT Africa (2011)

Table 5.5 Averaging over columns (per cent)

Mechanism	Averaging over columns (per cent)	Ranking
Sending or receiving e-mail	83.1	1
Finding or checking a fact	74.8	2
Social networking	71.3	3
Posting information or instant messaging	69.3	4
Downloading movies, etc.	66	5
Getting information for school or university work	61.4	6
Reading or downloading online newspaper or magazines, etc.	60.1	7
Playing or downloading video or computer games	59.6	8
Getting information related to health	58.7	9
Education or learning	57.4	10
Collaborating online	54.1	11
Getting information on goods	53.7	12
Getting information from government organisations	50.6	13
Looking for free education	49.9	14
Downloading software	46.7	15
Interacting with government organisations	32.7	16
Participating in distance learning	31.2	17
Telephoning over VoIP	30.6	18
Purchasing or ordering goods and services	23.1	19
Internet banking	19.2	20

Note. VoIP = Voice over Internet Protocol

about goods and services. The information contained in Table 5.4 can be summarised and made easier to interpret by aggregating over rows and columns, beginning with the latter in Table 5.5.

Use patterns and explanations

Table 5.5 shows the result of averaging across columns. It records a ranking of the most intensively used mechanisms covered by the sample. At 83.1 per cent, e-mailing is on average the most common of the uses to which the Internet is put in the sample countries. This is probably not surprising, since e-mail can be used for communication as well as seeking out information. The second most important mechanism, by contrast, is solely concerned with gathering information, involved as it is with finding or checking facts.

Perhaps the most striking aspect of Table 5.5, however, is the role played by entertainment-related mechanisms of use. In particular, four such mechanisms appear in the top ten entries in the table: namely, social networking, downloading movies, reading an online newspaper and playing video games. On first appearance, the prominence of entertainment-related uses may seem somewhat paradoxical, inasmuch as this form of use is seen as the preserve of those with high incomes and advanced Internet skills. But on further reflection, there do appear to be some alternative explanations for this finding.

One of them begins with the recognitions that entertainment is a relatively time-intensive activity and that time is a comparatively abundant resource in developing countries, especially among those with the lowest income levels. Becker (1965) for example, assumes that 'the value of an hour equals average hourly earnings' and thus that there is a 'one-to-one correspondence between earnings and the value of time'. Time is thus much more valuable in rich as compared to poor countries, a consideration which helps to explain the variegated efforts (such as frozen foods and microwave ovens) made in the former to economise on this factor. In poor countries, by contrast, production and use tend to be far more time-intensive (e.g., in cooking).

A second reason has to do with the age structure in Internet use, and in this regard one should bear in mind that according to the International Telecommunications Union (ITU), 45 per cent of users of this technology are drawn from those of less than 25 years of age. This percentage is almost certainly higher for entertainment-related uses, which, in the form of say, computer games, relies heavily on a relatively young audience. The point is then that in Africa (as most elsewhere in the developing world) the population is biased strongly toward younger ages. In 2009, for example, 43 per cent of Ethiopia's population and 45 per cent of Nigeria's were under the age of 15 (Todaro and Smith, 2011). In Europe, on the other hand, only about 15 per cent of the population is under this age.

The previous discussion suggests that it may be worthwhile to compare Internet use patterns in rich and poor countries. To this end, Table 5.6 presents the top ten uses in the United States and Kenya (a country from the sample that was arbitrarily selected for the comparison).

Apart from the appearance of e-mail in first or second position in the two countries, there is not much overlap between the entries in the two columns. A major difference seems to be that in the United States, the Internet is not as predominantly used for entertainment as in Kenya. In the former country, this technology is used mainly for getting information. Ironically, therefore, the country most in need of improved information is getting less of it. In a strictly neoclassical welfare framework, however, one cannot make interpersonal judgements about preferences even if in some cases the government may have to (e.g., in education).[5] In such cases, policy-makers might well conclude that entertainment is a less developmental activity compared to providing information on health, nutrition and so on.

Averaging across rows

The results of averaging across rows are shown in Table 5.7 that contains a ranking of all eleven countries according to the mechanisms that have already been described. This information conveys welfare significance because it tells us

Table 5.6 Ranking of top ten mechanisms: United States and Kenya

Kenya (per cent)		United States (per cent)	
Sending/receiving e-mail	89.1	Using a search engine	9.1
Social networking or video sharing	82.6	Sending or receiving e-mail	8.8
Posting information or instant messaging	80.8	Looking for information about hobby	8.4
Finding or checking a fact	77	Checking the weather	8.1
Downloading movies, etc.	76.2	Looking for information about a good or service you are thinking of buying	7.8
Playing or downloading video or computer games	71.8	Getting news	7.8
Getting information related to health	69.8	Going online just for fun or passing the time	7.4
Looking for free education	64.9	Buying a product	7.1
Reading or downloading online newspaper or magazines, etc.	64.8	Visiting a government website	6.7
Getting information for school-or university-related work/researching a topic	63.4	Using a social networking site like Facebook	6.7

Source: PEW Internet (2012) and Research ICT Africa (2011)

Note: PEW research bears no responsibility for interpretations presented or conclusions reached based on analysis of the data.

about the inequality of use between countries; whether, for example, it is more or less equal than the inequality of countries according to adoption. In one case, it would ameliorate such inequality, and in the other case worsen the problem. Even a glance at Table 5.7, however, is enough to indicate that there is no strong relationship between per capita income and Internet use, for there are poor countries (such as Uganda) near the top of the list and rich ones (such as South Africa) near the bottom (see a list of countries ranked by income per capita in Table 5.2). This impression of randomness is more rigorously confirmed by correlation analysis that finds a low correlation coefficient between income and Internet use (see Table 5.2).

Other promising determinants of Internet use are computer skills and computer literacy. After all, some uses are more difficult than others and the requisite skills may not be taught at either primary or secondary levels of the education system. In any event, moreover, 'not knowing how to' is the main constraint given by respondents for non-use of the Internet. Following Schmidt and Stork (2008), tertiary enrolment was chosen as the relevant proxy variable, but here again the correlation is low and difficult to explain. What does seem to correlate reasonably well (.54) with use, however, is the height of broadband Internet access tariffs, and unlike the others, this result fully conforms to what one would expect. The question of what determines price becomes an important and complex one. At least in the case at one extreme end of the country experience (see

Table 5.7 Averaging over rows (per cent)

Country	Average of Rows (per cent)	Ranking
Namibia	65.6	1
Uganda	61.7	2
Ghana	60.7	3
Nigeria	58.8	4
Kenya	52.9	5
Botswana	52.7	6
Tanzania	51.4	7
Cameroon	49.0	8
South Africa	45.5	9
Rwanda	43.5	10
Ethiopia	34.5	11

Source: Based on Research ICT Africa (2011).

Table 5.2), however, the answer seems relatively clear. I am referring to Ethiopia, and the explanation is to be found in the regulatory situation of that country. In particular,

> Ethiopia has Africa's last big telecoms monopoly. The absence of competition has seen a country of more than 80 m lag badly behind in an industry that has generally burgeoned alongside economic growth. . . .
>
> Ethiopia's authoritarian leaders are as keen as any on the economic benefits of modern telecoms but fear the political ramifications.
>
> (The Economist, August 24, 2013)

At the other extreme, Namibia has seen a falling price of Internet access due to favourable developments in the policy and regulatory environment.

A macro welfare adjustment

The data presented thus far are lacking in a macro dimension (i.e. in information at the economy-wide level). For that purpose, what also needs to be known is the percentage of Internet users at the macro level. For example, say 50 per cent of users are involved in using this technology to contact friends and family. What also needs to be known, however, is the percentage of users of the Internet that actually exist in the economy. Assume that figure is also 50 per cent. Then, 25 per cent of the adult population as a whole use the Internet to contact friends and family and thereby increase social capital.

Such a procedure can be extended to all countries in the sample as shown in Table 5.8. Perhaps the most striking effect of adding the percentage of users of the Internet to the first row of the table can be seen by comparing Uganda and Kenya.

Table 5.8 A Macro Welfare Adjustment

	Uganda	Kenya	Tanzania	Rwanda	Ethiopia	Ghana	Cameroon	Nigeria	Namibia	South Africa	Botswana
Per cent using family and friends to increase social capital	89.3	87.1	57.8	52.7	65.6	68.7	9.9	58.2	75.7	72.5	66.3
Per cent using the Internet	7.9	26.3	3.5	6.0	2.7	12.7	14.1	18.4	16.2	33.7	29.0
Macro Internet use of friends and family (per cent)	7.1	22.9	2.0	3.2	1.8	8.7	11.3	10.7	12.3	24.4	19.2
Country ranking	8	2	10	9	11	7	5	6	4	1	3

Source: Research ICT Africa

Note: Ranking is from highest to lowest.

Both countries show much the same percentage of those using family and friends to increase social capital. But Kenya enjoys a much higher percentage of the population (26.3) that is actually using the Internet. As a result, Kenya has a much higher position than Uganda after, rather than before, the macro adjustment has been made.

The impact of Internet use on social capital

The survey whose data are being used collects information specifically on the effect of Internet use on social capital. This is relevant to welfare, since many people feel that such an effect will be positive and hence that welfare will be increased. According to Quan-Haase and Wellman (2004), for example, the Internet can be thought of as an extra means of communications to foster extant social relationships and to follow norms of civic engagement. For the World Bank (2008), 'Information technology directly lessens the costs associated with imperfect competition. In this way, information has the potential to increase social capital-and in particular bridging social capital which connects actors to resources, relationships and information beyond their immediate environment'. In these ways, the participants will tend to benefit from the increased social capital thus achieved (James, 2009). Table 5.9 sets out the findings from the survey regarding the impact of Internet use on various dimensions of social capital.

Unlike those in the rest of the chapter, the results shown in Table 5.9 bear quite a striking resemblance to what was found for mobile phones using the same survey in a separate study (James, 2014). For one thing, in both cases, family is the most important vehicle through which the Internet affects social capital. Again like mobile phones, moreover, there is a preponderance of neighbouring East African countries (Uganda, Kenya and Tanzania) among the leading group of four. On the other hand, two countries from this region (Ethiopia and Rwanda) occupy the last two places, so it is difficult to talk of an 'East Africa' effect. This is clearly another area where further research is needed. Such research would do well to recognise that in other cases involving the social impact of the Internet, Uganda and Kenya also distinguish themselves in a positive sense. According to a South African technology research firm, for example, companies in Uganda and Kenya lead the way in blogging, interactions and online advertisements on Facebook, Twitter, MySpace and other social media (Kisakye, 2010).

Although these countries have similar scores on the first row, Uganda has a much lower score on the national percentage using the Internet, and the outcome shows an overall score for this country that is only one-third of Kenya's. On the other hand, South Africa and Botswana score best on Internet use, and this changes their overall performance in a favourable (for them) direction. Given the relatively high-income levels of these two countries, an inegalitarian tendency is imparted at the macro level to the initial result of using the Internet for contacting friends and family.[6]

Table 5.9 Internet use and social capital

Form of social capital	Country											
	Uganda	Kenya	Tanzania	Rwanda	Ethiopia	Ghana	Cameroon	Nigeria	Namibia	South Africa	Botswana	Average columns
Does your use of the Internet increase your contact with:												
• Those who share your hobbies	67.3	77.5	56.3	75.4	63.0	65.6	58.6	57.2	72.6	50.9	63.8	64.6
• Those who share your political views	42.4	30.8	36.8	2.6	11.4	31.1	14.0	40.3	19.0	24.0	35.5	26.2
• Those who share your religious beliefs	68.0	48.9	59.5	18.2	29.9	59.6	48.3	50.9	27.2	40.6	44.2	45.0
• Friends and family	89.3	87.1	57.8	52.7	65.6	68.7	79.9	58.2	75.7	72.5	66.3	70.3
• Colleagues	83.9	72.4	64.7	73.6	40.9	73.4	54.9	49.6	53.1	58.1	61.5	62.3
Average rows	70.2	63.3	55.0	44.5	42.1	59.7	51.1	51.2	49.5	49.5	54.3	
Rank	1	2	4	10	11	3	7	6	8	8	5	

Source: Research ICT Africa

Constraints on Internet use

In this final section, data that were collected in the survey on constraints to Internet use in the countries concerned are employed. One important question is whether these data tend to confirm or question the previous finding that the Internet access price correlates fairly closely with the use of this technology. Table 5.10 shows five barriers that are thought to constrain Internet use in all eleven countries, with a ranking for each one (1 denotes the best performance in a column, and 5 the worst).

By quite a wide margin, expense shows up as the dominant constraint, followed by the speed of the Internet and then the three more sociological variables, such as local content and lack of people to communicate with (the last-mentioned factor, one should note, stands in sharp contrast to the experience with mobile phones, where a shortage of people to communicate with was not readily apparent). The dominance of the price constraint serves to confirm the finding noted above of a moderate inverse correlation between Internet access tariffs and the use to which the technology is put in the sample. Note, though, that the correlation is far from perfect. On the one hand, there are countries such as Uganda, which is plagued by high Internet tariffs but manages to achieve the second highest use score. On the other hand, there are countries such as South Africa, which enjoys the lowest price but comes third from last on the use score. Explaining outliers such as these would seem like a profitable area for future research.

The seriousness of the price constraint becomes easier to understand when a comparison is made between African and other-country levels of this variable. In particular,

> For the few who accessed ICT networks at home – or more frequently in the workplace, educational institutions, or cybercafes – their usage was constrained by the high costs of communications, not least of all as a result of the high cost of international bandwidth. The average retail price for basic broadband in sub-Saharan Africa is US $ 366, per Mbps/m, compared to US $ 40 in Europe and India.
>
> (Gillwald, 2010: 82)

From a policy point of view, therefore, attention should be most obviously paid to lowering prices, especially in countries where these are high relative to others in the sample. What seems clear, though, is that a purely economic approach to this issue would not suffice.[7] As Gillwald rightly puts it:

> adopting stronger political economy approaches to policy reform, approaches that analyze the interaction of the state and the market to facilitate the implementation of reform, would go some way toward explaining why the reform paradigm, which promotes competition, dilutes incumbents' monopoly-wielding power, and adopts universal service policies that

Table 5.10 Constraints on Internet use

Use constraint	Country										
	Uganda	Kenya	Tanzania	Rwanda	Ethiopia	Ghana	Cameroon	Nigeria	Namibia	South Africa	Botswana
Lack of interesting content	5	5	5	5	5	5	5	5	5	4	3
Lack of local language content	5	4	4	4	4	3	4	4	4	5	5
Too slow	3	3	3	2	2	1	3	1	2	2	2
Too expensive	1	1	1	3	1	2	1	2	1	1	1
Too few people to interact with	2	2	2	1	3	4	2	3	3	3	4

Source: Based on Research ICT Africa

do not distort the incentives of the market actors, has only been adopted piecemeal, limiting its efficacy. Such an approach would lay *greater emphasis on the role of institutions* and the interplay among them. This would allow a greater understanding of what enables or inhibits their effectiveness to translate a market-based reform paradigm into action.

(Gillwald, 2010: 80, emphasis added)[8]

It is worth noting that a similar conclusion was reached in the lengthy debate over appropriate technology for developing countries. While simple economic logic seemed to dictate the use of labour-intensive technology in labour-abundant, capital-scarce economies, actual practice rarely jelled with this prescription, and even state-owned enterprises tended to choose in favour of relatively complex, capital-intensive alternatives. Part of the reason, it appeared, had to do with the dominance of political over economic logic (Stewart, 1983).

Conclusion

Using a detailed survey of Internet use in eleven African countries, this chapter has explored the implications for welfare of moving from a consumption theory of ownership and utility to one based on use and benefits. To a large extent, the implications were explained in terms of the analysis of a basic data set, with the eleven countries shown on the one axis and Internet use mechanisms on the other. This showed not only the leading use mechanisms across countries, but also the leading country beneficiaries across mechanisms. As far as the former are concerned, the outcome is that a high percentage of Internet use is concerned with entertainment as compared to a selected developed country whose inhabitants seemed more concerned with obtaining basic information of various kinds. The pattern of leading countries, however, proved more difficult to interpret, as it displayed virtually no correlation with the familiar variables of income and education. The only reasonable correlation was achieved in the case of broadband Internet access tariffs, placing the explanatory focus firmly on telecom markets and regulation. This is clearly an area where considerable further research is needed.

I also examined the effect of Internet use on social capital – that is, on whether and to what extent social ties were enhanced between the relevant parties. The results obtained in this regard are very similar to those calculated in the same way for mobile phones in a separate article using the same survey. In both cases, for example, the interpersonal relationship most influenced by information technology was between family and friends. Again, like mobile phones, moreover, the leading countries were drawn almost overwhelmingly from East Africa. The reasons for this however are not clear and open up yet another area for further research. More generally, the results presented in this chapter should be interpreted less as firm conclusions and more as a tentative future research agenda.

Notes

1 See, for a review of the mobile phone studies, James (2013).
2 Research ICT Africa is a South Africa-based organisation that conducts research on ICT policy and regulation (see researchICTafrica.net). I am grateful to this institution for granting me access to their data set. The underlying method is available at ReasearchICTAfrica.net ('Household, Small Business and Public Institutional e-Access and Usage Survey', 2011). The results of the survey are available on request from the author. The questionnaire itself is available at ResearchICTAfrica.net.
3 In my scheme, the distribution of benefits constitutes an important part of welfare. Many studies show that relative use is an essential part of how people value consumption.
4 See James (2009).
5 In education, for example, governments frequently have to confront the issue of rationing scarce Internet time to different groups, and therefore have to formulate criteria for such a purpose.
6 In the case of mobile phones, by contrast, the relatively poor countries score well on overall use, which favours equality at the macro level.
7 A promising approach at the micro level is to bring low-cost, off-the-grid broadband access to rural Kenya and hopefully to other African countries.

> The technology making the project possible is called dynamic spectrum analysis, which enables wireless devices to opportunistically tap into unused radio spectrum in the television frequency band, as well as solar-powered based stations. . . . As television has begun to switch from analog to digital around the world, even more of this spectrum can be used to fulfill those needs.
>
> (Garnett and Otieno, 2013)

8 Recall also the earlier citation from *The Economist*, which emphasises the role of politics in the Ethiopian telecom market.

References

Becker, G. (1965). A theory of the allocation of time, *The Economic Journal*, 75:493–517. Doi:10.2307/2228949.

Garnett, P. and Otieno, L. (2013). *Bringing Low-Cost, Off-the-Grid Broadband Access to Rural Kenya, Microsoft on the Issues*. Available at http://blogs.technet.com/b/microsoft_on_the_issues/archive.

Gillwald, A. (2010). The poverty of ICT, policy research and practice in Africa, *Information Technologies and International Development*, 6:77–87.

James, J. (2009). Sharing mechanisms for information technology in developing countries: Social capital and the quality of life, *Social Indicators Research*, 94:43–59. doi.org/10.1007/511205-008-9335-3.

James, J. (2014). Product use and welfare: The case of mobile phones in Africa, *Telematics and Informatics*, 31:356–363. doi.org/10.1016/j.tele.2013.08.007.

Kisakye, J. (2010). *Ugandans Reap from Faster Internet*, London: The Observer.

PEW (2012). PEW Internet and American Life Project, trend data, Washington, DC.

Quan-Haase, A. and Wellman, B. (2004). *How Does the Internet Affect Social Capital?*, Cambridge, MA: MIT Press.

Research ICT Africa (2011). *Internet Data*. Available at www.researchICTAfrica.net.

Schmidt, J. and Stork, C. (2008). Towards evidence based ICT policy and regulation: E-skills, *Research ICT Africa*, 1, Policy Paper 3.

Sen, A. (1985). *Commodities and Capabilities*, Amsterdam, The Netherlands: North Holland.

Stewart, F. (1983). Macro-policies for appropriate technology: An introductory classification, *International Labour Review*, 122:279–293.

Todaro, M. and Smith, S. (2011). *Economic Development*, New York, NY: Addison-Wesley.

World Bank (2008). *Social Capital and Information Technology*. Available at http://web.world bank.org/WBSITE/EXTERNAL/TOPICS/EXTSOCIALDEVELOPMENT.

6 Institutional and societal innovations in IT for developing countries

In the developed countries, access to information technology (specifically mobile phones and the Internet) takes place almost entirely by means of individual ownership and use. Application of this model in the developing countries, however, would exclude the vast numbers with neither the incomes nor the skills to operate the technology effectively (though the relatively rich minority in developing countries would be included among the beneficiaries). Fortunately, there are a plethora of local innovations in these countries which deviate from developed-country institutions. In this chapter, I describe some of these innovations, which tend to be locally generated, low-cost and reliant on sharing of different kinds. In some cases, the innovations have had a discernible macro impact on the domestic economy. They emphasise what can be achieved when local innovation is directed towards the rural poor rather than the average conditions in developed countries (see also Chapter 2).

My goal is to reveal the assumptions underlying the developed-country model of adopting IT (information technology) and to contrast them with an alternative model, based mostly on institutions and innovations in the developing countries themselves. The alternative set of assumptions in my view offers a far more compelling way to reach the rural poor in less-privileged parts of the world. The common feature of this alternative model is that it involves mainly societal and institutional change. The crucial point is that such behaviour is not part of some platform for IT diffusion in the future. Rather, it is becoming increasingly recognised as change that can and often does affect a relatively large part of an economy. Indeed, when sharing of mobile phones is taken into account, the digital divide seems, on recent evidence, to all but disappear (see below).

First, however, let me describe the nature of innovation that is based in and occurs for institutional and other conditions in the developed world. Our approach, in common with much of the study of development in general, makes a simple binary distinction between rich and poor countries. This, of course, has its limitations. As stated by Todaro and Smith, for example,

> The simple division of the world into developed and developing countries is sometimes useful for analytical purposes. Many development models apply across a wide range of developing country income levels. However, the wide

income range of the latter serves as an early warning for us not to over generalize. Indeed, the economic differences between low-income countries in sub-Saharan Africa and South Asia and upper-middle-income countries in East Asia and Latin America can be even more profound than those between high-income OECD and upper-middle-income developing countries.

(Todaro and Smith, 2011: 41)

As a practical matter, the arbitrariness is resolved by definitions laid out by large international institutions such as the IMF and World Bank which are widely accepted by member countries. These institutions also seek to ameliorate the problem by defining four income groups rather than just rich and poor. Few people, in any case, would deny that there are enough similar characteristics among the poor countries to distinguish them as a separate category from the rich parts of the world.

The nature of innovation in developed countries

Singer (1970) and Stewart (1977) and others have emphasised how technical change in the rich countries is designed to fit in with the socio-economic situation that prevails there at any given point of time. I am referring here to infrastructure, skills and income and institutions, among other circumstances. This would not be problematic if research was also being done on the very different problems in the developing world. Unfortunately, however, less than 15 per cent of global R&D is directed at the problems of the poor, and little of even that goes towards the conditions of poor rather than rich persons in developing countries (though this percentage has risen with the rapid development of India and China). There are, of course, exceptions to this general rule, such as the low-cost laptop developed by MIT,[1] but they are relatively few and far between.

Given what has just been argued, it follows that IT innovations from rich countries match closely with the conditions in those countries. So, for example, the vast majority can afford a mobile phone, a computer and an Internet connection and have the skills and infrastructure that are needed (e.g., electricity). That is to say, the use of IT is obtained here via individual ownership in the vast majority of cases, and from individual use the benefits are obtained.

For developing countries, on the other hand, low income levels act as a strong deterrent to ownership, especially in the case of the Internet, but also to some extent mobile phones as well. Some other means of access need to be found instead. Like some of the other models in this chapter, they have much to do with sharing IT.

Local innovations in developing countries

Local innovations differ from developed-country innovation systems by breaking the link between ownership and benefits and between use and benefits. In the developed-country model, benefits accrue from ownership and use of a technology that is designed for and suited to the conditions prevailing in such countries. Local innovations, by contrast, are adapted to the conditions prevailing in the poorer parts of developing countries i.e., poor, rural areas or marginalised parts of the urban sector. Local innovations tend to arise spontaneously rather than over the time it takes to perform R&D. Relatedly, these new technologies tend to be relatively simple, involving societal or *institutional* change rather than change in products and processes. Consider in this regard the case of Grameen telecom which is discussed below. The mobile phone used in the programme is *not* the main innovation, which consists rather, in the way access to the technology is organised in villages. *This shift towards institutional innovations rather than physical innovations marks an important change in the earlier literature on appropriate products and processes.* That earlier debate (as already described) tended to focus on the choice of technology, and in particular on the selection between capital- and labour-intensive methods of production (see also the cases of technological blending described below).

Sharing mobile phones

It has long been recognised that some amount of non-commercial sharing of mobile phones takes place among owners, family and friends in developing countries. What is new is first the recognition that such behaviour is often a major phenomenon in those countries (in that it involves a relatively high proportion of the population). James (2011) points to a 'culture of sharing' in many parts of the developing world, and uses available micro survey data to estimate how many people use but do not own a mobile phone. The greater the number of such people, the more do conventional estimates overstate the true size of the divide between rich and poor people. For example, if a developing country has a penetration rate of 16–17 per cent, then a sharing rate of five users per owner would be enough to provide access to some 100 per cent of the population (these numbers by the way are drawn from several African countries).

Paid forms of sharing range from street vendors in Africa to the well known case of Grameen Telecom in Bangladesh and other countries (such as Uganda) to which it has spread. The Grameen case is one in which a female member of the bank is appointed to act as the mobile phone operator in the village. Aggregating over all villages in the country means that tens of millions of clients now have access to a phone. Furthermore, the case is one of the most detailed attempts to measure the impact on the poor. Bayes et al. (1999), for example, have sought to measure the welfare effect of the phone on the poor. In particular, they have studied the concept of consumer surplus (CS), which is defined as 'the difference between the price consumers actually pay and the price they would be willing to pay' (Bayes et al., 1999: 29). The authors find that the presence of village phones provides all strata of the population 'with a fair amount of CS'. But it is the poor who gain the highest CS from the village phones, while the non-poor derive the lowest (as reported also in other chapters).

Table 6.1 Cybercafé use, selected countries

Country	Cybercafé total	Cybercafé per capita
Argentina	18,500	462,5
Brazil	58,000	302,1
Peru	32,000	1.142,9
Uganda	25,400	819,4
Moldova	474	118,5
Sri Lanka	1,000	50
Mongolia	105	35

Source: Coward et al. (2008)

Sharing the Internet

Sharing the Internet often takes place in cybercafés which offer communal access to the technology. Table 6.1 shows the total and per capita number of cybercafés in a selected group of developing countries.

In absolute terms, Brazil has by far the largest number of cybercafés, but because it has such a large population, the per capita availability falls below the leading countries. Peru is in fact the highest of those countries, and its superior performance is undoubtedly due largely to the so-called cabinas publicas. These institutions were introduced in 1998 by a group of entrepreneurs seeking to spread the Internet more widely, on the one hand, and to make profits, on the other. Although they are located in urban areas in and around Lima, the cabinas serve mostly low-income neighbourhoods. Such is the popularity of these sharing institutions that in some cases they have begun to exert a macro level influence on IT activity ('Honduras and Peru reported that more than half of all ICT usage in these countries takes place at cybercafés' (Coward et al., 2008)). It is not yet clear, however, how much the poor have been positively affected by these institutions in the developing world as a whole (see below).

More recent research on public access facilities includes work by Clark and Gomez (2011). They find that 'digital literacy of staff and local relevance of content may be more important than fees in determining user preference for public access venues. These findings are important to public libraries, which tend to offer free services, but where perceptions of digital literacy of staff and locally relevant content tend to be lowest, compared to telecentres and cybercafés' (Clark and Gomez, 2011: 1). In the same year, a study by Gomez and Baron asked whether public access computing (telecentres, cybercafés and libraries) contributes to community development. Using several case studies they find that 'for the most part, there is little or no connection between the new information and communications opportunities people have gained through PAC [public access computing], and the community development needs in their contexts' (Gomez and Baron, 2011: 9).

One of the most important findings of this line of research is that there is a clear concentration of public access venues located in urban areas. While

telecentres have a high proportion of non-urban locations, public libraries and cybercafés are primarily urban, with 64% and 91% respectively, in
urban locations. Furthermore, on average, only 31% of the public libraries
offer ICT as part of their services and these libraries tend to be in urban
centres. Given that cybercafés account for 73% of all public access venues
studied (the majority in urban areas), and given that over half the public
libraries are urban, it is clear that public access to ICT is mostly an urban
phenomenon. With a concentration in urban areas and populations, public access to ICT, for the most part, fails to serve the majority of the rural
population in the countries studied. The urban/non-urban divide is by far
the most significant divide in public access to ICT.

(Clark and Gomez, 2012: 1)

And since the poor in most developing countries mostly reside in rural areas,
they are generally untouched by the institution of public access venues.

Sharing institutions without use

In the 'cabinas publicas', benefits from the Internet are generated by individual
use of this technology. But in many cases, perhaps especially in Africa, the lack
of skills that are required to use the Internet are simply not available. It is here
that the sharing without use model comes into play (note in this regard that, as
reported by Schmidt and Stork (2008), there are numerous African countries in
which lack of skills is regarded as the most important reason for not using the
Internet). A key figure in this model is the intermediary who comes between
the beneficiaries and the technology, someone who is familiar with the technology and the community.

There are numerous instances of innovation with these properties: Internet
kiosks in India are one of the most prominent examples.[2] Some are being targeted
directly towards the alleviation of rural poverty. The Sustainable Access in Rural
India (SARI) is a case in point. At its peak, SARI comprised more than 80 rural
kiosks in the Tamil Nadu state and provided, among other services, e-government.

The most popular e-government service provided at these centres was
income and community (caste) certifications. Income certification verifies
that the individual is below the Indian poverty line while community certification confirms that they are members of a historically under-privileged
community. Receiving these government certifications allows an individual
to avail of various government schemes and welfare programmes.

(Best and Kenny, 2009)

One is entitled to speculate that this local innovation has tended to benefit the
poor more than richer groups in the area. Note that in this and other examples
of Internet kiosks, the computer is operated by an intermediary rather than the
ultimate beneficiary.

Other 'share without use' models can be described as examples of technological blending, where older technologies are combined with IT. An ingenious case in point is the combination of radio and the Internet employed in Kothmale in Sri Lanka (and also in Nepal). The Kothmale project was designed by a local broadcaster for the many in that country with access to radio. The Internet is brought to these people by means of radio broadcasts, which invite local doctors, teachers and lawyers to discuss and interpret findings that are relevant for the local population. Another example of blending takes place in Pondicherry, India, where a loudspeaker is used to bring weather information (gleaned from the Internet) to local fishermen.

It is unfortunately not possible to gauge the extent of the 'share without use' model in developing countries. But given the pronounced lack of Internet skills in these countries, perhaps specially in Africa, what can be said is that the model has vast potential to bring the benefits of the Internet to the uneducated rural masses.

Sharing mobile phones vs public access centres

I have discussed above numerous models of innovation in IT for developing countries. An important question concerns the relative success of these models in reaching the population in these countries (with particular reference to the poorest groups). In this section, I use what data are available to compare (shared) mobile phones and public access centres such as telecentres, cybercafés and libraries.

In Ethiopia, for example, my calculations show that the total number of households which own a mobile phone equals 0.7 million. I also know from survey data that each such household shares with five persons. These numbers suggest that there were roughly thirty-four million nonowning users in the country. There are few studies which provide data on the number of users per public access centre against which to compare the number of nonowning users of mobile phones, but from a survey of another African country it appears that only some twenty persons per day visit a sample telecentre (James, 2011). Since there are relatively few telecentres in Africa, even aggregating up to the national level would not diminish the conclusion that the sharing of mobile phones has a relatively large impact compared to at least one form of public access. It is difficult to imagine a different conclusion with the other public access centres since, as Coward et al. (2008) have shown, they are relatively scarce in developing countries.

Innovation systems and product quality

Repeated innovations in and for the developed countries mean that the digital divide is constantly widening. For example, increases in bandwidth and 3G mobile phones have enlarged the gap just when it has fallen considerably prior to the quality increases (especially in the case of mobile phones). These newer technologies well suit the needs of the majority of those in the rich countries

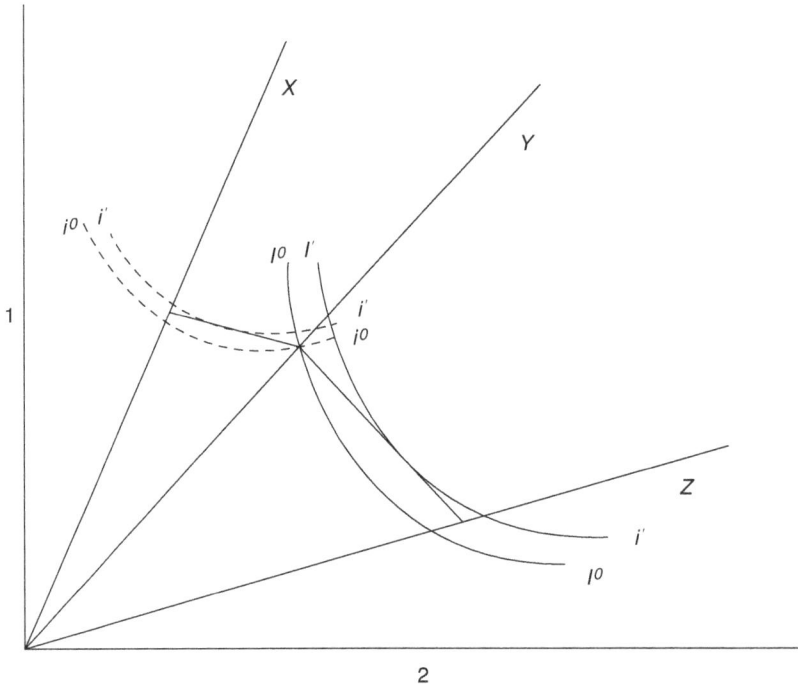

Figure 6.1 Quality in innovation systems

and a small group of persons in poor countries, but they are too expensive for the majority of those living in the latter countries. Consider the situation portrayed in Figure 6.1.

On the axes are simple functions requiring limited bandwidth and sophisticated functions requiring far more bandwidth. Here too, this binary distinction is over-simplified, and the reality is that there will be some amount of heterogeneity within the groups. However, the difficulty of finding a 'perfect' cut-off point does not mean we should not try. A relatively clear distinction may be between basic (or starting-level) Internet vs higher functions found primarily in rich countries.

From the origin I have drawn 2 lines representing two products. The one, Y, is intensive in simple functions and the other is intensive in more sophisticated functions. The indifference curves represent average preferences in the two countries. i1 i1 belongs to the developed countries and I1 I1 to the developing parts of the world. The new product X moves the former countries to a higher indifference curve at B. The latter countries, however, remain at A because they are not willing or able to pay for the increase in product quality. An important policy point arises from this analysis: namely, that if product standards are set too high, it will simply exclude the majority of people in developing countries from gaining the benefits of IT. A more pressing task is to enable the rural population in those countries to gain access to the basic functionings (given endemic skills and other constraints).

Conclusion

In the developed countries, innovation in IT is based heavily on R&D and is closely geared to local conditions such as income, infrastructure, institutions and culture. It is directed, moreover, almost entirely to solving rich-country rather than poor-country problems. But application of this formal model to the latter countries would exclude the vast majority of the population with neither the income nor the skills required to operate some of this technology. Fortunately, another model has evolved for the poor countries which is low-cost and based on various forms of sharing that greatly enhance the access to IT on the part of those with low incomes. It is perhaps best exemplified by the sharing of mobile phones in developing countries, and for the Internet by various examples of technological blending. What can be described as local innovations often serve the interest of the rural population and those with relatively low incomes. In some cases, these 'appropriate' innovations have had a macro – as opposed to village-level – impact on the domestic economy. They illustrate what can be achieved when local innovation is introduced in a poor rather than rich-country environment. Sharing mobile phones in various ways seems to be an immensely more promising avenue than public access centres (such as telecentres, cybercafés and libraries) in the spread of IT across the developing world.[3]

Notes

1 The so-called One Laptop per Child programme is discussed and criticised by James (2010) and Warschauer and Ames (2010).
2 Other prominent examples include e-Seva, n-Logue and Gyandoot.
3 There are though differences within institutions (Gould and Gomez, 2010).

References

Bayes, A., von Braun, J. and Akhter, R. (1999). *Village Pay Phones and Poverty Reduction*, Bonn: ZEF.

Best, M. and Kenny, C. (2009). ICTs, enterprise and development, in T. Unwin (ed.) *ICT4D, Information and Communication Technology for Development*, Cambridge: Cambridge University Press, 177–205.

Clark, M. and Gomez, R. (2011). The negligible role of fees as a barrier to public access computing in developing Countries, *Electronic Journal of Information Systems in Developing Countries*, 46. Doi:10.1002/j.1681-4835.2011.tb00323.x.

Clark, M. and Gomez, R. (2012). Libraries, telecenters and cybercafés: A comparison of different types of public access venues, in M. Clark and R. Gomez (eds.) *Libraries, Telecenters and Cybercafés: A Comparison of Different Types of Public Access Venues*, IGI Global.

Coward, C., Gomez, R. and Ambitar, R. (2008). *Libraries, telecentres and cybercafes: A study of public access venues around the world.* Submission to IFLA, Quebec.

Gomez, R. and Baron, L. (2011). Does public access computing really contribute to community development? Lessons from libraries, telecentres and cybercafes in Colombia, *Electronic Journal of Information Systems in Developing Countries*, 49,1:11. Doi:10.1002/j.1681-4835.2011.tb00346.x.

Gould, E. and Gomez, R. (2010). New challenges for libraries in the information age: A comparative study of ICT in public libraries in 25 countries, *Information Development*, 26,2:166–176. doi.org/10.1177/0266666910367739.

James, J. (2010). New technology in developing countries: A critique of the one-laptop-per-child programme, *Social Science Computer Review*, 28,3:381–390. doi.org/10.1177/0894439309346398.

James, J. (2011). Sharing mobile phones in developing countries: Implications for the digital divide, *Technological Forecasting and Social Change*, 78,4:729–735. doi.org/10.1016/j.techfore.2010.11.008.

Schmidt, J. and Stork, C. (2008). Towards evidence based ICT policy and regulation: E-skills, *Research ICT Africa.net*, 1, Policy Paper 3.

Singer, H. (1970). Dualism revisited: A new approach to problems of the dual society, *Journal of Development Studies*, 7,1:60–75. doi.org/10.1080/00220387008421348.

Stewart, F. (1977). *Technology and Underdevelopment*, London: Macmillan.

Todaro, M. and Smith, S. (2011). *Economic Development*, 11th edn., Boston: Addison-Wesley.

Warschauer, M. and Ames, M. (2010). Can one laptop per child save the world's poor?, *Journal of International Affairs*, Fall–Winter, 64,1:35–51. Available at www.jstor.org/stable/24385184.

7 The macroeconomic consequences of the One Laptop per Child program

With the goal of bringing a low-cost computer to every child at the primary school level in developing countries, the One Laptop per Child (OLPC) programme has been widely discussed. By now, there is considerable literature on various aspects of the proposal, for example, Warschauer and Ames (2010) and James (2011). This literature suffers, however, from two main weaknesses. One of them is that very little economic reasoning has been used in the debate, especially of a macro kind. This is not least problematic because there are already some countries (Peru and Uruguay) that have adopted OLPC computers (known as 'XO') at a national level and one that is planning to do so (Rwanda). The other weakness is that while there is much discussion about whether the program is 'good' or 'bad', very little attention has been paid to the question of when it might be appropriate.

In this brief chapter, I seek, at least as a beginning, to redress these weaknesses, and thereby to point the literature in a different and more policy-relevant direction.

The basic ratio

In most countries where they are used, XO computers are confined to the micro level of pilot projects. However, there are already two countries, Peru and Uruguay, that have adopted the OLPC project in full. The former, for example, has provided such computers to about one million primary school students. Under these conditions, the impact of the project can no longer (only) be assessed from a microeconomic perspective; macroeconomic considerations also need to be taken into account. What is at issue here is the ease with which the OLPC programme can be absorbed into the education budget.

The basic ratio, R, may be defined for a country as the number of primary school students × cost of a laptop ($200) divided by the amount of the GDP spent on education. (A similar paper of mine uses this ratio in the context of India and it asks whether the complete introduction of OLPC there would be worth it. It was submitted later but published earlier than this chapter. See Current Science 106(8): 1061–1063.)

Table 7.1 Peru, Uruguay and Rwanda

	Population	Primary school children	GDP per capita	Education budget	R
Peru	29.8 million	1 million	$10600	$81bn	$200m/0.81bn = 25$ per cent
Uruguay	3.3 million	395000	$15900	$1.9bn	$79m/1.9bn = 42$ per cent
Rwanda	11.46 million	1 million	$552.5	$7.1bn	$200m/298m = 67$ per cent

Source: Nationmaster, UNESCO, World Bank and own calculations

Peru, Uruguay and Rwanda

Judging by the data shown in Table 7.1, it is clear that the values of R in Peru and Uruguay diverge quite sharply from one another. In particular, assuming a price per computer of $200, at 25 per cent the size of the education budget, the ratio defined in the previous text is almost six times larger in Peru as it is in Uruguay (where it is less than 5 per cent). The difference in the numerator between the two countries is primarily because of population size. As shown in Table 7.1, the former has a substantially larger population than the latter, and consequently a substantially higher absolute number of primary school attendees.

An explanation of the difference in the denominator between the two countries has to concern itself with the relatively high expenditure on education in Uruguay as compared to Peru. The latter, for example, spent only 2.6 per cent of its GDP on education in 2011, as compared to 4.5 per cent for the former in the same year (World Bank). Moreover, an explanation of this difference should draw attention to differences in teacher salaries because these typically form the largest share of the education budget as a whole. Recall from Table 7.1, for example, that Peru is about 50 per cent less well off than Uruguay, and the difference between them in teacher salaries may very well reflect this.

I turn next to a much poorer African country, Rwanda, which has committed itself to a complete adoption of the OLPC model in the future. Consider first the numerator of the ratio R, namely the cost of providing each primary school pupil with a laptop computer. Assuming that there will then be one million such pupils (OLPC, 2013), the endeavour is likely to cost the government about $200 million. Using World Bank data, the total education budget is calculated as 4.2 per cent of the GDP in that country, or 4.2 per cent of 7.1bn (equal to some $298 million).

Dividing the numerator by the denominator gives a percentage equal to 67. In other words, two-thirds of the education budget would be devoted to laptops for primary school children, manifestly a non-viable state of affairs (see more on this below).

Rich and poor countries

The analysis is now extended somewhat by comparing poor (low-income) developing countries with countries at the other end of the spectrum, that is, those that are rich and developed. As far as the former are concerned, the study of Rwanda has already raised the severe difficulties likely to be faced by poor African societies in fully implementing the OLPC idea.

In developed countries, by contrast, where absolute spending on schools, teachers and other items of the educational budget tends to be relatively high, the ratio R is usually lower and the OLPC project is correspondingly easier to accommodate. In the UK, for instance, I have calculated that the value of R would be less than 1 per cent (using government data of 822 million primary students and education spending equal to £88.6 billion). Moreover, while there is thus clearly no problem there of accommodating the full OLPC program, the question is whether it makes any sense to do so where there are already 1.9 or 1.8 students per computer in primary schools (The Guardian, January 13, 2012).

Implications of budgetary imbalance

Of the three countries in which the OLPC has been or intends to become fully implemented, two of them, Peru and Rwanda, have suffered or will suffer from an acute imbalance in their education budgets. It is telling to note in this regard that the expenditure on laptop computers in the Peru case could instead have been spent on more than 18,000 primary school teachers (UNESCO, 2012). Indeed, it is the shortage of these very resources that is said by some to underlie the limited educational achievements of the OLPC in that country (IDB, 2012).

Note, finally, that the calculations made in the preceding text substantially underestimate the actual costs of implementing the OLPC proposal at the national level, for I have only counted the costs associated with an increased supply of computers. Nevertheless, in reality, governments progressing beyond the pilot phase (which may be at least partly be financed by foreign aid) will also have to pay for the sizeable amount of training that is needed to impart the requisite skills to teachers using computers as an education tool. In addition, and relatedly, governments will need to provide new software as well as repair/ maintenance capabilities. All these costs, moreover, have to be incurred against expected scholastic benefits, which, at this point, seem highly uncertain. Thus, I can conclude that while the technology is in many ways suited to African conditions, the institutional practice of providing each pupil with his or her own laptop is in many cases not appropriate.

References

IDB (2012). Technology and child Development: Evidence from the One Laptop per Child Program, Working Paper 304.

James, J. (2011). Low-cost computers for education in developing countries, *Social Indicators Research*, 103,3:399–408.

OLPC (2013). *Rwanda Laptop Project*. Available at http://blog.laptop.org/tag/olpc-rwanda/.

UNESCO (2012). Institute of Statistics, Paris.

Warschauer, M. and Ames, M. (2010). Can one laptop per child save the world's poor?, *Journal of International Affairs*, 64,1:33–51.

8 Sharing mechanisms for IT in developing countries, social capital and quality of life

Sharing information technology (IT) is not much of an issue in the developed countries, since most of the individuals there derive their benefits from the technology via ownership. In the case of mobile phones, for example, inhabitants of these countries typically have at least one subscription.[1] In developing countries, by contrast, lower average incomes mean that ownership is limited to a much smaller group of individuals.[2] For the rest of the inhabitants of these countries, therefore, sharing of one kind or another may be one of the only means of gaining access to and benefitting from IT. So far, however, no-one has looked at 'IT for development' specifically from the standpoint of sharing, though there are already many forms of this type of behaviour. My task accordingly is to fill this rather yawning gap in the literature by providing an analytical classification and comparison of the different forms of sharing across the three main types of IT: namely, the Internet, computers and mobile phones.[3] For part of this task I use the concept of social capital, which according to many people will be increased by IT. According to Quan-Haase and Wellman (2004), for example, the Internet can be regarded as an additional means of communication to facilitate existing social relationships and to follow patterns of civic engagement.

As such, this technology tends to increase existing patterns of social contact and civic engagement. For the World Bank, 'Information technology directly lessens the costs associated with imperfect information. In this way, information has the potential to increase social capital-and in particular bridging social capital which connects actors to resources, relationships and information beyond their immediate environment' (World Bank, 2008). In all of these ways the participants will tend to benefit from the increased social capital thus achieved. The World Bank also provides an example of how this general mechanism has actually worked in practice. Thus,

> Goods can now be sold via the Internet which permits access to greater markets which before could only be reached by those with enough capital to afford transportation. Cooperatives or craftspeople [who share the Internet] are beginning to sell their wares to consumers in industralized countries via the Internet. This typically requires an Internet accessible non-governmental organisation (NGO) to act as intermediary between the

producers and the consumers . . . In this example the internet offers opportunities to enhance social capital among craftspeople within a cooperative and builds bridging social capital by connecting producers and consumers who would otherwise not be able to do business together.

(World Bank, 2008)

Other examples involving the relationship between IT and social capital will be cited below.

The limits of ownership-based access to information technology

At the heart of my argument in this section lies the important notion of a technological system. According to Stewart (1977),

> There are technical linkages between different parts of the system which mean that much of technology comes as a package which cannot be separated and introduced bit-by-bit, but as a package which goes together.
>
> The requirements of a technique extend beyond the material inputs directly involved in the productive process to managerial inputs and infrastructural services. Thus the efficient use of a particular technique . . . may impose particular demands for energy, water and transport . . . Levels of living of the labour force may be another technical requirement. The required labour input, in terms of energy, concentration, punctuality and literacy are related to the technology . . . This is not to argue that each technique imposes a unique set of requirements, and can only be operated if these requirements are met. But any variation tends to lead to variations in the productivity of the techniques and *sufficient deviation from the sort of inputs for which the techniques are designed may lead to a total breakdown.* For example, cars designed for advanced-country-roads will not work where there are no roads at all and the length of life and efficiency of operation will be seriously affected by the different conditions with roads that are not tarmacked and with few mechanics.

(Stewart, 1977: 7)

Since most innovations emanate in and for developed countries they tend to fit in with the characteristics of the technological system in those countries, such as high average incomes, advanced educational attainments and skills, a highly developed infrastructure and so on. In developing countries, by contrast, only a relatively small part of the economy exhibits the features of this kind of modern technological system. It is typically that part of the economy whose residents live in urban areas, and which in other respects approximates the features of a rich-country system. A high percentage of the IT that is adopted by a poor country is concentrated in this so-called formal sector of the dual economy and is used on the basis of individual ownership.

A few examples help to illustrate these points. Consider first the situation with respect to mobile phones (whose relatively low price and lack of user skill requirements make them easier to own than the Internet or computers). According to Donner (2005: 2), it is true of the poorest countries that 'mobile ownership is still mainly for the privileged middle class and elites in urban areas. For many others, the costs of mobile ownership and use remain prohibitively high'. More recently, however, even in rural areas, ownership and use have grown relatively rapidly, though there are still 20 per cent of the population in developing countries who do not use mobiles. With respect to the Internet, the problem will tend to be even more pronounced, partly because the technology itself is much more expensive, and partly because the skills needed to operate this technology are rather formidable. (It has long been recognised, for example, that there is something distinctive about the skill requirements for information technology as compared to other innovations (van Dijk, 2005)). Indeed, there is clear evidence that recent growth in Internet use has taken place predominantly among members of the formal sector in developing countries (though again, this tendency has been ameliorated in recent years). A survey conducted in three African countries, for example, finds that 'level of education was . . . a major factor influencing propensity to use e-mail and internet. Those with secondary and or post-secondary education were far (more) likely to use internet than other educational tiers' (Gamos, 2003: 41).

These privileged groups in developing countries, however, amount to only a small percentage of the population as a whole, and cannot thus long serve as the market for future Internet growth (the exact length of time depends on factors such as the level of per capita incomes, the degree of inequality and the speed of adoption in each case). This key point has been made by the authors of a case study on the diffusion of IT in South Africa, when they conclude that

> household Internet penetration is concerningly low at 3.5% of the respondents, with most Internet users acquiring access at work or school. *With the low levels of household ownership* (12%) and the high cost of the fixed line infrastructure *there is the danger that the Internet market will rapidly reach saturation.*
>
> (Gillwald, 2005: 250, emphasis added)

More generally, it is estimated that in Africa as a whole, some 80 per cent of the population owns neither a mobile nor fixed-line telephone.

The problem then becomes the limits to adoption of IT in the so-called traditional technology system which co-exists with its modern counterpart described in the previous paragraphs. The former system is characterised by relatively low levels of technology, skills and infrastructure, and it is located mainly in the rural sector of the economy. It is worth emphasising that much of the production in the traditional technological system is based on non-capitalist modes of production, such as non-wage labour, where sharing of output is customary. Production units tend to be small-scale and dispersed.[4] As already noted, there are limits to which IT can be adopted via ownership in the modern sector as well as the traditional sector. Much of the task in gaining future access to IT in developing countries will thus have to be about sharing mechanisms of one kind or another with particular reference to the traditional technology system. In the next section, I present a basic classification of sharing mechanisms with reference to the Internet, the computer and the mobile phone. We will have reason to observe that many of the mechanisms described do not appear in calculations of the digital divide or the quality of life in developing countries.

Conceptual dichotomies and the basic classification

Even in the developed countries, there are those who do not own IT and need to rely on some alternative form of sharing access, such as cybercafés, computer rental after school hours and so on. These same mechanisms also exist (mostly in the formal sector) in developing countries. As I will show in this section, however, many sharing mechanisms are designed to fit in specifically with the traditional technology systems prevailing in the rural sectors of developing countries. The classification scheme presented below thus comprises elements from both modern and traditional technology systems in developing countries. For each type of IT, the actual elements in the schema are derived from the use of a conceptual dichotomy and a number of sub-indexes (Table 8.1). The traditional, as opposed to the modern, mode of sharing is reflected in the dichotomy as well as the sub-categories.

Let us first consider how these factors operate in the case of the Internet (shown in column 1 of Table 8.1).

Table 8.1 shows that as far as the Internet is concerned, the relevant dichotomy is thought to be between users and non-users – or, in other words, between those who have the (many) skills required to operate the technology effectively and those who do not.[5] The distinction is important, because those in the latter

Table 8.1 The basic classification of sharing mechanisms

The Internet	Mobile phones	Computers
(a) *Users* • Inside institution with connection (e.g. telecentres) • Sharing a connection outside the institution	(a) *Non-commercial* • Family, friends • Beeping	(a) Individuals benefit from sharing a remote computer
(b) *Non-users* • At a distance • Close range	(b) *Commercial* • Micro-finance institutions • Other sharing institutions	(b) Individuals benefit as part of a communal institution • Institutional change in sharing institution • Technical change to promote sharing in institutions

Source: Partly based on James (2007)

category need to be able to benefit from the Internet without actually using it. In developing, as opposed to developed, countries, the share of non-users to users is relatively high. It is for this reason that many of the examples below are devoted to sharing mechanisms for non-users of the Internet. This group, I should emphasise, typically contains individuals with the lowest incomes in developing countries, for at the one extreme are the richest groups whose incomes enable them to own an Internet-enabled computer. Somewhat less wealthy are the middle to high-middle groups which possess the skills needed to operate the Internet. Because of their relatively advanced user capabilities, these groups will tend to earn more than the larger number of individuals without such skills at the lowest levels of the income distribution.

Within the category of users, I have made a distinction between sharing that takes place within an Internet-connected institution and the case of sharing a connection by institutions that do not have already have one (though they do possess computers). The latter type of sharing is important, for example, in the case of schools in developing countries, since, as shown in Table 8.2 for a selected group of such countries, the percentage connection to the Internet is rather low as compared with schools in developed countries with perfect or near-perfect percentage connections. More recent data compiled by the ITU (2014) show that connectivity in schools in developing countries remains abysmally low. In particular, the results indicate that fewer than 10 per cent of schools in Latin America, the Caribbean, Asia and Africa were actually connected.

As regards non-users of the Internet, I have made a further distinction between distance and face-to-face intermediation. In both cases, the benefits are derived by the intervention of an intermediary who comes between the technology and the community. But in the one case, the intermediation takes place at close

Table 8.2 Percentage schools connected to the Internet in a selected sample of countries

Developed countries (2005)	Per cent
Belgium	93
Japan	99
Sweden	99
United Kingdom	99
USA	100
Ghana	1
Malawi	1
Costa Rica	1
Trinidad and Tobago	15
Jamaica	10
Mongolia	10

Source: World Bank, ICT at-a-glance tables

Note: I could find no more recent data.

range (with rural Internet kiosks for example), and the other at a distance (as in the case of community radio stations that use the Internet for the benefit of their listeners). Either way, though, sharing of a high technology, the Internet, takes place very much in the context of a 'low' technology system with limited incomes, skills and infrastructure.

The dichotomy chosen for mobile phones (see Table 8.1) also reflects a distinction between traditional and modern sharing systems in developing countries. In particular, much of the sharing of this technology in the former system (unlike the latter) takes the form of free use within a household or between friends. This is partly a matter of culture. Many authors speak of a 'culture of sharing in developing countries'. For instance 'the Philippines has a culture in which household members generally share resources' (Pertierra, 2005). Or again, in a different context, 'Although a mobile phone may nominally belong to a single person, in some African countries it is regarded as the property of the community, because there is a culture of sharing the tools of communication' (Lopez, 2000).

There is also a greater need to freely share mobile phones within households and between friends in developing countries. On the one hand, household size tends to be large in developing as opposed to developed countries. Within the former, family size tends to be greater within rural than urban areas, reflecting the traditional aspects of rural areas. Over time in the developed countries, household size has fallen from around 5.5 to 2.6 over the past century or so (Bongaarts, 2001). By the late 1990s, the average for many regions in the developing world was still between five and six (Bongaarts, 2001). On the other hand, large family size often tends to be associated with higher poverty,[6] thus making it more difficult for individual family members to buy their own mobile phones. Although country-comparative data on sharing mechanisms that do not involve payment are unavailable, household survey data do indicate that many owners lend out their phone on a wide scale. In Botswana, for instance, 62.1 per cent of the phone owners share this technology with their family, 43.8 per cent with their friends and 20 per cent share their phone also with their neighbours (Sebusang et al., 2005).

Note that the term 'non-commercial' also applies whenever users of mobile phones employ the device of 'beeping' one another. For instance, beeping once may mean that the person is at a certain meeting point, whereas beeping twice might signify that there is a half-an-hour delay. However, one beep might also be a signal for saying 'hello'. In these ways, using a mobile phone effectively becomes costless.[7]

All other sharing devices for mobiles shown in Table 8.1, however, are commercial in the sense that they have to be paid for, albeit in very small units. I have singled out micro-finance-based phone services partly because they operate in a different way from other sharing mechanisms and partly because of the unusual degree of success they have enjoyed, primarily in the form of Grameen Telecom:

> As an extension of the original Grameen Bank endeavour to make small group loans to its (female) members, the idea of the Telecom project is to lend money to a Bank member in each village in Bangladesh for the

purpose of purchasing a mobile phone. The phone owner then sells call-time to the other villagers, who, it seems are willing to pay a relatively high proportion of their incomes on this service.

(James, 2007: 290)

Note that the intensity of sharing in this project is heightened by the notification of villagers of any incoming calls. This service effectively provides participants with an important element of phone ownership, namely, the ability to receive incoming calls.

The final form of IT considered in Table 8.1 is the computer, which may or may not be used as a component of Internet access. As noted in Table 8.2, for example, relatively few schools in developing countries have an Internet connection, but many of them will have at least one computer. Using computers in a basic way is more demanding than using a mobile telephone but less exacting than navigating the Internet. The dichotomy I have chosen for computers captures the idea that sharing can take place either by individuals who access a computer at a remote location, or by individuals within a sharing location itself. In the latter case, there is a sub-division into institutional and technical determinants of the extent to which a computer is actually shared within, say, a school.

Examples classified by category

As described above, Table 8.1 has yielded numerous categories within which to classify different forms of sharing information technology in developing countries. The purpose of this section is to use these categories to classify actual examples of how IT is shared in the Third World. My hope is that this exercise will provide some analytical structure to what is now just a fragmented list of cases.

The Internet

Because they are relatively numerous, it is useful to remind ourselves of the different categories for the Internet by using the tree-diagrammatic form in Figure 8.1

Of the four sharing mechanisms shown in Figure 8.1, let me begin with case 1, where benefits are extracted in communal organisations such as telecentres, schools and firms, much as they also are in developed countries. Recall from Table 8.2 for the case of schools, however, that the amount of sharing in this category will be proportionally lower in developing than developed countries. Note too that even for those who do engage in sharing the Internet in communal institutions, the actual benefits are likely to be much lower in the former countries. Rural telecentres, for example, are subject to power cuts, repair and maintenance problems and limited bandwidth (Etta and Wamahiu, 2003). Inserting just the modern technology into a traditional technology system usually leads, as noted above, to a much less than desired outcome (James, 2005b). More recent research on Tanzania, moreover, indicates that telecentres are used predominantly by the younger generation.

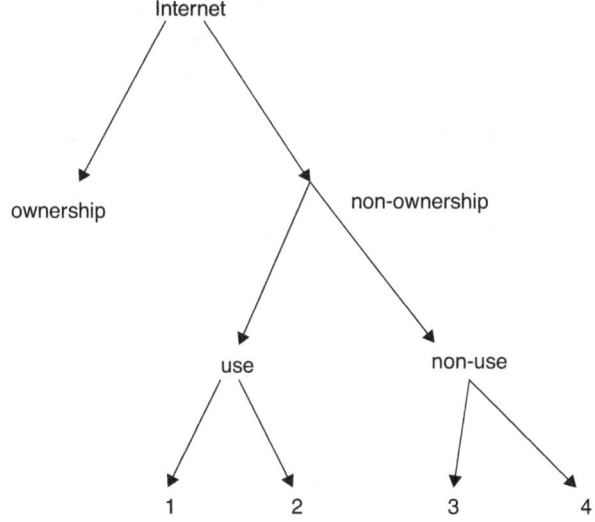

Figure 8.1 Mechanisms for sharing the Internet in developing countries
Source: The author

The second category of sharing by users is rather different from the first, and more development-oriented (with particular reference to rural areas in poor countries). It involves the sharing of an Internet connection by rural institutions that do not already have one and which are prepared to use this technology in non-synchronous ways (that is, in cases where some delay is involved in the delivery of information as compared with the real-time alternative used in developed countries).[8]

Consider first in this regard South Africa's 'Wizzy Digital Courier', which was designed specifically for the rural parts of the school system and works even in the many isolated schools that have no telephone line. The system is based instead on the physical movement of data between schools such as these and a location which does have an Internet connection. In particular,

> It is a system that involves transporting data saved on a USB memory stick back and forth between one central computer . . . and unconnected machines in outlying areas . . . Using a program . . . designed with open-source software, the school can compose e-mail messages and specify topics for an Internet search. Then it can send a teacher or a gardener off – by bicycle, perhaps – to place the memory stick into a central computer miles away, upload the current batch of communiqués, and retrieve the results of yesterday's requests.
>
> (James, 2008: 177)

By cutting the costs of information in this way, villagers are here benefitting from the increased social capital induced by the introduction of the Internet. The gains accrue in the form of a better education for students.

'DAKNET' is a similar, and arguably the most prominent, example of non-synchronous Internet connectivity for developing countries. This example has been succinctly described by its founders in the following way:

> As an implementation of very low cost asynchronous ICT infrastructure we have developed a store-and-forward wireless network for rural connectivity. . . .
>
> The DakNet wireless network takes advantage of existing communications and transportation infrastructure to distribute digital connectivity to outlying villages lacking digital communications infrastructure. DakNet combines physical means of transportation with wireless data transfer in order to extend the Internet connectivity provided by a central uplink or Hub (eg. A cybercafé, VSAT, or post office) to kiosks in surrounding villages.
>
> Instead of trying to relay data over a long distance (which can be expensive), DakNet transmits data over short point-to-point links between kiosks and portable storage devices called Mobile Access Points (MAPs). Mounted on and powered by a bus, motorcycle, or even bicycle, the MAP physically transports data . . . between kiosks and a Hub (for non-real-time Internet access) . . . By employing short-distance radio links, DakNet allows for small low-cost low-power radio devices to be used.
>
> (Pentland et al., 2004)

It is worth noting at this point that both these examples of rural Internet sharing were specifically designed to fit in with local conditions, not the least of which is that time is a relatively abundant resource which should be used intensively rather than economised upon.[9] Both cases, however, are limited to the relatively small number of individuals who can actually operate the Internet effectively. The next category deals with bringing the benefit of this technology to the vast number of individuals lacking those skills (see the second mechanism in Figure 8.1)

Non-user beneficiaries

As noted above, sharing in this subcategory occurs with the help of an intermediary who comes between the Internet and the rural beneficiaries. In the one case, intermediation takes place at a distance, and in the other at close range. Let me deal first with the former.

It is perhaps best exemplified by a technology 'blending' project in Pondicherry, India, that was designed by the M.S. Swaminathan Research Foundation. What occurs is that ocean wave reports are gathered from the Internet and communicated four times a day to the mostly poor fishermen in the village. Such information helps them to decide whether and when to ply their trade each day. I have referred to this as a case of technological blending because a recent technology (the Internet) is combined with a much less sophisticated and older technology (namely, the loudspeaker that is used to broadcast the weather forecast to villagers).

Another type of blending occurs when the Internet is combined with radios belonging to some of the poorest groups in rural areas of developing countries. One of the best-known examples of how this type of blend can yield benefits for the local community is the Kothmale Community Radio in Sri Lanka. I am referring here to the way in which this project uses a novel 'radio-browsing' technique to bring together those who transmit information from the Internet and those who receive it. In particular,

> The daily programmes respond to queries from listeners. Presenters first select relevant, reliable websites and broadcast the programme with local resource persons as studio guests (e.g. doctors for a health programme) who discuss the contents of the mostly English-language sites directly in the national languages. They also describe the websites and explain how they are browsing from one page to another. Thus, listeners not only get the information they requested, but they understand how it is made available on the web. They can respond to the programme, and they know that essential data will remain available in the community database if they wish to make individual use of it. With this daily radio programme, there is continuity within a common learning process encouraging greater inter-activity with and by the community.
>
> (Hughes, 2003: 3)

Use of the Internet in this case thus enhanced social capital by facilitating more intensive and favourable relationships among those who listen to the radio broadcasts, the presenters and those who appear as voluntary experts on particular topics. The presenters, for example, learnt about tea production in other countries and, after checking with the experts, shared some of the information with his listeners by overcoming market imperfections in gaining knowledge. There were also spontaneous off-the-air relationships between the listeners and the experts that were promoted by the radio broadcasts (such as discussions 'in the street') (Hughes, 2003).

It is not clear if Radio Kothmale continues to provide 'Internet-browsing' programmes, but very similar models can be found in Nepal and Bolivia.[10]

None of these cases, however, would be relevant if the benefits derived from the Internet by rural residents involve codified transactions, of one sort or another, that are determined by specific individual needs. Whereas the distance approach is geared towards general information and a wide audience, the face-to-face approach is based on customised transactions of various kinds. This is the reason why so many rural Internet kiosks have emerged to serve the needs of poor, illiterate rural villagers. Many of the services supplied by local or regional government offices involve birth and death certificates. Prior to the introduction of the Internet, these documents would need to be obtained by long and costly visits to the nearest government offices, involving, as they did, standing in long lines and paying bribes along the way. With an Internet connection and an intermediary housed in a rural kiosk, however, these costs of imperfect information can be greatly reduced and social capital increased (in the manner described above by the World Bank).

It is difficult to find a better illustration of how such kiosks can benefit farmers than the 'Bhoomi' Project in India:

> Bhoomi was launched by the Karnataka state government in order to computerise land records on a massive scale, and to make them available to farmers for a small fee. No fewer than 20 million records have been computerised, and in the form of computerised kiosks, can be securely called up. *Such records are crucial to farmers because they are used for other purposes such as loan requests.*
>
> (James, 2005a: 117, emphasis added)

Other services provided by rural Internet kiosks include help with registration of complaints and submission of applications for issuing certificates and loans. Market price data of agricultural crops in different locations can also be bought for a small fee (I have taken these examples from the 'Gyandoot' project in Madhya Pradesh).

Note that the advantages of an Internet kiosk for rural inhabitants in developing countries can also be delivered by a mobile vehicle as opposed to a fixed stand. In the 'Computers on Wheels' (COW) project in India, for example, mobile 'information providers' use specially equipped motorcycles to travel between villages.[11] The cycles are equipped with, among other things, an Internet-connected laptop computer and are designed to access villages without passable roads. The answers to specific questions posed by villagers are returned once the information provider has established an Internet connection somewhere else. As such, this project also falls into the category of non-synchronous information supply as described above.

Note too that intermediaries, who bring the benefits of the Internet to non-users on a one-to-one basis, need not themselves be located in rural kiosks. Indeed, in the interesting case of 'Babajob.com' in India, what is being sold is the entry on the Internet of a poor job-seeker's profile, and this occurs not in rural areas but in a large Indian city. Specifically, the prospective workers dictate their personal data to a secretary who enters them into a computer. Then, once his or her photograph has been taken, the resulting profile appears on the Internet. The idea is that matches take place through 'friend of-a-friend' networks (Giridharadas, 2007). Imagine for example that two employers are friends and that one of them needs a chauffeur. He can then look up the page of his friend's chauffeur and see which of the chauffeur's friends are looking for work. In this way 'Babajob' reflects online the underlying process by which hiring decisions are made in real life by Indians: 'using chains of personal connections'. The Internet offers another form of conduct for these key relationships, one that may at times be more efficient than the existing alternatives. Either way, though, the technology would seem to increase the social capital of the community.

Mobile phones

In spite of the expectations expressed earlier about the likelihood of widespread sharing of mobile phones in (especially poor) developing countries, there are remarkably few data on this topic. Some of the most detailed evidence has been collected by Stork, in his field study of Namibia (2005). What he found was that,

> Respondents tend to share their mobiles with family first, friends second and neighbours third. Nearly 30% of respondents regularly share their mobiles with other family members, compared to about 16% with friends.
>
> The percentage of respondents that shares their mobiles with family, friends or neighbours is distinguished by household income. One can clearly observe a trend of higher income households tending to share their mobiles less, which can also be attributed to more mobiles being owned per household.
>
> . . . It can be observed that respondents living in rural areas are more willing to share their mobiles with others than respondents in major urban or other areas . . . Interestingly, only 2.77% of respondents that share their mobile phones charge friends, family or neighbours a fee for the use of their cell phones.
>
> (Stork, 2005: 112–113)

A different study by Goodman (2005) specifically studied the impact of mobile phones on social capital in two other African countries, South Africa and Tanzania. He finds that 'In both countries there was a high degree of sharing mobiles for free with friends and family (and sometimes for money). This indicates that mobiles may be acting as a social amenity, a tool to be shared and a focus for social activity, as well as a tool for communications' (Goodman, 2005: 63). 'Mobiles were thus enabling people to invest in and draw on social capital' (Goodman, 2005: 66).

If sharing of this kind does not thus seem to be merely a minor phenomenon (at least in these African countries), it is also the case that more people (especially in rural areas) benefit from it than cross-country data on mobile subscribers would suggest, and to this extent, the digital divide as conventionally measured will tend to be overstated. The same may hold true for another form of non-commercial sharing known as beeping.

Consider, for example, the case where the sender is already using some form of communal facility such as a telecentre. He or she may then costlessly beep recipients with a variety of different messages. Donner reports research conducted by a consultancy firm Gamos which concluded that '38% of users of public payphones and telecentres in Uganda, Botswana, and Ghana regularly beeped mobile users from these phones' (Donner, 2007). Donner's own research suggests that beeping tends to amplify and strengthen existing relationships and in this way to promote social capital in the relevant communities (2007).

I turn next to the category of sharing mobile phones that was referred to as 'commercial' in Table 8.1. In the sub-sections of this category, a distinction is drawn between micro-finance-based schemes with particular reference to Grameen Telecom and other sharing arrangements. The case for treating this initiative separately is based on several recognitions. One of them is that Grameen Telecom can be regarded as a pioneer sharing project in focusing on the creation of micro-enterprises based on IT, but it is also the first such facility to foster village phones on the basis of digital, wireless telephones. As a private rural telecommunications project, Grameen is also a pioneer in targeting poor women in villages as operators (and owners) of the phone. Indeed, social capital in this project is most often discussed in connection with these women who become empowered by owning a phone and become leaders of the community with new ties to its members (Stanley, 2005).

But the uniqueness of this project probably rests most heavily on the phenomenal success it has had in the rural areas of Bangladesh and to a lesser extent in Uganda and certain other developing countries. Unlike many other attempts to use IT for the benefit of the rural poor, the effect of Grameen Telecom extends well beyond the level of a village or region. Indeed, data taken from the official site of 'Grameenphone' indicate that 67,000 villages in Bangladesh are affected by the project. If one assumes, as others have done (Cohen, 2001), that a single phone covers some 70 rural inhabitants, then no fewer than forty-five million villagers in that country have access to a mobile phone, thanks to the Telecom initiative.[12] Data from the same source suggest that there are some 6.3 million borrowers (village phone operators), 97 per cent of whom are women. Although it is not known how many of the forty-five million villagers with access to a mobile phone actually make use of one, it is clear that the number will run into the tens of millions. It also bears emphasis that whatever this number happens to be, it will not be registered under the number of mobile subscribers in international data on Bangladesh. Here too, therefore, one can talk of an overstatement of the digital divide. Quality of life gains from the project include the difference before and after the introduction of phones. 'The cost of a trip to the city ranges from two to eight times the cost of a single phone call, meaning that the real savings for poor rural people of between $2.70 and $10 for individual calls' (Stanley, 2005: 3).

Beyond the Grameen model are sharing initiatives that originate on the supply and demand sides of the transaction. With regard to the former, for example, the activities of 'Vodacom' in South Africa bear mention. In particular, that firm, as part of its licensing agreement, has generated a Community Service shared access model which resembles an Internet cafe. Local entrepreneurs re-sell phone time from specially converted containers that are connected to the mobile phone network. This shared-access model is said to account for a significant proportion of mobile calls made within South Africa. In Uganda, local operators act as intermediaries in much the same way as the examples described above in relation to the Internet. The need arises because not even the very limited skills required for using the mobile phone can always be taken for granted.

In some cases, for instance, potential users are illiterate and require the operator to enter the desired number and confirm it before making the call. Yet another mechanism on the supply side is analogous to the case of 'computers on wheels' described above in relation to the Internet in India. In particular, in West Bengal rural postmen provide mobile phones at the doorstep of villagers in some 12,000 villages. This scheme is known as 'Grameen Sanchar Sewak' and it was initiated in 2002 by the Indian government (see, for example, www.india.gov/hindi/sectors/communications4php).

On the demand side, one example is the pooling of resources to purchase phone time by a peer group of one kind or another. This makes sense when no individual alone can afford to buy the smallest available amount of phone time. By means of pooling resources, however, the group is able to purchase that minimum amount which may then be loaded onto one individual's phone and shared by all the members. In the same spirit is the case where people who cannot afford a mobile phone use prepaid cards to make calls from a handset that belongs to someone else. Again, a report on Kenya concludes that 'Phone sharing is also common. Some users just buy a SIM card (which is plugged into the phone to grant access to the network) for less than a dollar and borrow a phone from someone else' (Business Week, Special Report, September 24, 2007). As in the case of South Africa and Tanzania noted above, sharing of phones in Kenya may well also be enabling people to invest in and draw upon social capital.

Computers

I have already referred to cases where the Internet has been 'blended' with more traditional forms of technology such as radios and loudspeakers. As we now turn to the last form of IT, computers, another form of blending comes immediately to the fore. In terms of Table 8.1, it falls under the category where individuals benefit from access to a remote computer (as will become apparent, this case, like the previous examples of blending, demonstrates that sharing IT is a serious means of reaching the groups with the least access to this form of technology).

I am referring specifically to 'Kisan Call Centres' in India, which operate on the basis of combining telephones and computers, and were initiated by the Department of Agriculture in 2004.[13] The project represents a novel application of IT to extension services in that country. With a national toll-free number, any farmer, literate or not, can call the Centre which is staffed by someone with an agricultural degree and is proficient in the local language (calls from within a given state are answered from a call centre located in that state). Such a person is the first to deal with an incoming call and enters information about the caller and the query into a computer. If the query cannot be answered at this level, it is referred by computer to a specialist in the area. Data collected in the computers from registration of caller queries are used as inputs to an information base. Kisan makes particular sense in a country like India with a relatively large number of public payphones. As yet, however, no data are available to measure the extent to which the centres are actually being used.

The remaining examples under the category of sharing computers occur within a communal institution rather than between an individual and a remote computer. I have drawn a distinction among these remaining examples between increased sharing that results from some kind of institutional change as opposed to a technological innovation. A school, for example, undergoes an institutional change when it allows its computers to be used by community members outside of regular hours. This is what occurred in a 'World Links'[14] project in Uganda where an after-school community telecentre project was developed (Best and McClay, 2001). 'Under this program, schools in rural Uganda that are equipped with computer labs and VSAT-based Internet connections are opening up their labs to outside clients in the afternoon and evenings on a cost-recovery basis' (Best and McClay, 2001: 81). 'World Links' continues to apply this basic model in a variety of other developing countries such as Zimbabwe, India and Brazil. Extended use of the Internet has of course the potential to increase social capital as noted above.

MIT's One Laptop per Child programme (OLPC)[15] also represents an institutional change, in that it seeks to drastically alter the existing number of children per computer in developing countries. Indeed, as its name suggests, the programme seeks to deliver a (low-cost) laptop to each and every child at school. One would have thought that so extreme an objective – of effectively precluding sharing in schools – was well justified by its promoters. Yet quite the opposite is the case, and the objective is scarcely defended at all, which is the more problematic because it runs directly counter to any notion of appropriate institutions in developing countries.[16] That is that developing countries cannot afford the standards set in the rich part of the world, and have therefore to set lower standards to incorporate more than a small percentage of the population.[17] Since the number of students per computer that is thought to be pedagogically effective in developed countries is thought to be four to five (US Department of Education, 2001), the OLPC is effectively setting an even *more* stringent rule. Not surprisingly, the application of this rule would severely imbalance the education budget in developing countries. Peru, for example, is said to be currently spending on laptops nearly a third of the education budget normally available for capital expenditures (Talbot, 2008).

Regardless of whether each child does receive a laptop, however, the OLPC product does embody a desirable technological feature that promotes sharing within a school, and as such exemplifies the subcategory devoted to these forms of innovation in Table 8.1. In particular,

> Using standard wireless protocols, the laptops are automatically able to form a 'mesh network' where each machine acts as both laptop and router, able to pass information between computers.
>
> If one laptop is switched on in range of an internet connection . . . all other laptops on the network can share the access . . . If there is no Internet access, the laptops can still share data, video and information through the mesh.
>
> (BBC, 2007)

No less interesting are the attempts to use multiple mice in schools in developing countries. The idea began with the familiar observation that in India there are insufficient computers in relation to the number of students. The result is that only the child with access to the mouse benefits from use of the computer (or at least the benefits of users exceed those of non-users of the mouse). The latter sometimes vie for control of the mouse and gradually lose interest and begin doing something else (Pawar et al., 2006). One way to overcome this problem, according to Pawar et al. (2006), is to provide each child with a mouse and cursor on the screen. Initial trials to develop software allow a single computer to connect with multiple mice, each of which belongs to an individual student. Thus, instead of giving each student an entire computer (as in the OLPC programme), this approach gives students the chance to share a *single* computer.

Conclusions

I began this chapter by discussing the relationship between IT, social capital and quality of life in developing countries. I took note of the considerable potential that IT has for increasing the social capital among communities, and thus for improving the quality of life for the inhabitants. The Internet, for example, can be regarded as an additional means of communication to facilitate existing social relations and to follow patterns of civil engagement. IT also has the ability to reduce the costs associated with imperfect information, and hence the ability to intensify social relationships beyond the level that would otherwise be the case. An example involving the use of IT to increase trade among a group of artisans shows that the connection between IT and social capital does actually occur in developing countries.

I then suggested that the institution of ownership which dominates the modes of access to IT in developed countries will not get us far in providing access to IT in the rural sector of developing countries (though this too is changing with the introduction of low-cost phones). It was emphasised that sharing provides the main viable alternative to individual ownership in this context. As yet, however, no one has looked at IT specifically from the standpoint of sharing. The existing examples remain unrelated to one another, and this fragmentation does not well serve those who want to see a more structured classification of the examples that do exist. Distilled from the examples in the text, Table 8.3 hopefully fills part of this gap for the Internet, mobile phones and computers.

Several aspects of Table 8.3 bear emphasis. One of them is that most of the examples cited therein originate in and for developing countries. They tend to exhibit properties that are different from what is found in developed countries (although a few examples are common to both parts of the world). Policy needs to figure out whether these indigenous innovations are cost-effective, and if so, how to replicate them (as has already been done for the case of Grameen Telecom). Relatedly, we need to recognise that several important instances of sharing

Table 8.3 A summary of sharing mechanisms with examples

The Internet	Mobile phones	Computers
(a) *Users*	(a) *Non-commercial*	(a) *Individuals benefit from sharing a remote computer*
• Inside institution with connection (telecentres, schools)	• Family, friends (Namibia case study)	• Kisan Call Centres, as blending computers and telephones
• Sharing a connection outside the institution (Wizzy Digital Courier: Daknet)	• Beeping (regular occurrence in Uganda, Botswana and Ghana)	
(b) *Non-users*	(b) *Commercial*	(b) *Individuals benefit as part of a communal institution*
• At a distance (blending project in Pondicherry; community radio in Sri Lanka)	• Micro-finance institutions (Grameen Telecom)	• Institutional change in sharing computers (using computers by community outside regular school hours; OLPC)
• Close range (rural Internet kiosks such as Bhoomi and Gyandoot)	• Other sharing institutions (Vodacom in South Africa; local intermediaries in Uganda; pooling resources; SIM card; mobile phones delivered to the doorstep of villagers in India)	• Technical change to promote sharing in institutions (OLPC; multiple mouse)

go unrecorded in official cross-country data, and to this extent the digital divide between rich and poor countries is overstated.

Finally, I have suggested that the much-vaunted OLPC project is unrealistic and wasteful because the recommended number of students per computer in developed countries is only four or five, not one-to-one. Spending an amount greater than would be required for four or five students can properly be regarded as unnecessary. The whole point of the appropriate technology idea is that developing countries need not adopt the same standard as the developed world, so even the goal of four to five students per computer is highly questionable. Something like ten to twelve students per computer probably makes much more sense (especially if combined with a scheme such as multiple mice).

Notes

1 According to World Bank ICT at-a-glance tables.
2 A useful set of case studies dealing with this point is contained in Gillwald (2005).
3 The main goal is not to endorse all the examples that I cite. It is rather to use these as representative of different types of sharing mechanisms.
4 For a full description of the technological aspects of the traditional sector, see Stewart (1977).
5 van Dijk (2005) has an extensive discussion of the skills required to use the Internet.
6 Often the causality has to do with economics. For women with relatively high salaries in the developed countries, the opportunity cost of having a child is much higher than for a woman who is poorly paid or entirely unemployed in a developing country.
7 For a case study of this phenomenon, see Donner (2007).

8 James (2008) argues that non-synchronous technologies are appropriate for developing countries since they are intensive in time (the abundant resource) and saving in capital (the scarce resource). They buy a lower capital cost by using more of time, which seems perfectly rational. Some examples of non-synchronous technology have been explicitly based on this recognition.

9 For a detailed discussion of this case and its cost compared to other forms of rural communication, see Pentland et al. (2004).

10 Note that community radio has the potential to reach hundreds or thousands of people with just one digital connection. As such, it stands as a potentially potent way of overcoming the digital divide

11 The project is described by the BBC News on 15 July 2002 under the heading 'Villagers try out net on wheels' (available at www.news.bbc.co.uk/i/hi/sci/tech/2124712.stm).

12 Grameenphone is available at www.grameenphone.com.

13 'Kisan Call Center' (available at www.manage.gov.in/kisan/default.htm).

14 The homepage of 'World Links' is available at www.world-links.org.

15 The homepage of OLPC is www.laptop.org.

16 On the concept and application of appropriate technology, see Stewart (1977).

17 If, for example, housing standards were set at developed-country levels, the vast majority of the population would be excluded. What are needed are intermediate standards that lie somewhere between having no standards and developed-country levels.

References

BBC News (2007). Factfile, XO Laptop. Available at http://news.bbc.co.uk/21hi/technology/6679431.stm. Accessed July 13, 2017.

Best, M. and McClay, C. (2001). Community access in rural areas: Solving the sustainability puzzle, *The Global Information Technology Report*, 2001–2.

Bongaarts, J. (2001). Fertility and reproductive preferences in post-transitional societies, in R. Bulatao and J. Casterline (eds.) *Global Fertility Transition*, New York: Population Council.

Cohen, N. (2001). What Works: Grameen Telecom's Village Phones, *World Resources Institute, Digital Dividend*. Available at http://digitaldividend.org/pdf/grameen.pdf.

Donner, J. (2005). *Research approaches to mobile use in the developing world: A review of the literature*. Paper Presented at Conference on Mobile Communication and Asian Modernities, City University of Hong Kong, June 7–8. Available at www.jonathandonner.com/donner-mobrev.pdf.

Donner, J. (2007). The rules of beeping: Exchanging messages via intentional "missed calls" on mobile phones, *Journal of Computer-Mediated Communication*, 13,1.

Etta, F. and Wamahiu, S. (2003). *Information and Communication Technologies for Development*, Senegal: Codesria, IDRC.

Gamos Ltd (2003). *Innovative Demand Models for Telecommunications*. Available at www.teleafrica.org/pdfs/FinalReport.pdf.

Gillwald, A. (ed.) (2005). *Towards an African e-Index: ICT Access and Usage*, Johannesburg, South Africa: The Link Centre, Wits University School of Public and Development Management.

Giridharadas, A. (2007). Poverty inspires technology workers to altruism, *New York Times*, October 30.

Goodman, J. (2005). Linking mobile phone ownership and use to social capital in rural South Africa and Tanzania, Vodafone Policy Papers, No. 2.

Hughes, S. (2003). Community multimedia centres: Creating digital opportunities for all, in B. Girard (ed.) *The One to Watch: Radio*, FAO, Rome: New ICTs and Interactivity.

ITU (2014). Measuring the Information Society Report, Geneva.

James, J. (2005a). Technological blending in the age of the internet: A developing country perspective, *Telecommunications Policy*, 29,4. Doi:10.1016/j.telpol.2004.11.010.

James, J. (2005b). The global digital divide in the internet: Developed country concepts and third world realities, *Journal of Information Science*, 31,2. Doi:10.1177/0165551505050788.

James, J. (2007). From origins to implications: Key aspects in the debate over the digital divide, *Journal of Information Technology*, 22:284–295.

James, J. (2008). Time-intensive information technology and human welfare in developing countries, *Prometheus*, 26,2:165–177. Doi:10.1080/0810902080202976.

Lopez, A. (2000). The south goes mobile, *UNESCO Courier*, July–August.

Pawar, U., Pal, J. and Toyama, K. (2006). *Multiple mice for computers in education in developing countries*. IEEE/ACM International Conference on Information and Communication Technologies for Development, ICTD.

Pentland, A., Hassan, A. and Fletcher, R. (2004). DakNet: Rethinking connectivity in developing nations, *IEEE Computers*, 37,1:4–9. Available at https://ieeexplore.ieee.org/document/1319279/.

Pertierra, R. (2005). Mobile phones, identity and discursive intimacy, *Human Technology*, 1,1:23–44.

Quan-Haase, A. and Wellman, B. (2004). How does the internet affect social capital?, in M. Huysman and V. Wulf (eds.) *Social Capital and Information Technology*, Cambridge: MIT Press.

Sebusang, S., Masupe, S. and Chumai, J. (2005). Botswana, in A. Gillwald (ed.) *Towards an African e-Index: ICT Access and Usage*, Johannesburg, South Africa: The Link Centre, Wits University School of Public and Development Management.

Stanley, R. (2005). *Village Phone: A Tool for Empowerment*, Grameen Foundation, USA Publication Series.

Stewart, F. (1977). *Technology and Underdevelopment*, London: Macmillan.

Stork, C. (2005). Namibia, in A. Gillwald (ed.) *Towards an African e-Index: ICT Access and Usage*. Johannesburg, South Africa: The Link Centre, Wits University School of Public and Development Management.

Talbot, D. (2008). OLPC laptop gets windows, *Technology Review*, May 16.

US Department of Education. (2001). *Internet Access in US Schools*, Fall.

van Dijk, J. (2005). *The Deepening Divide*, London: Sage.

World Bank (2008). *Social Capital and Information Technology*. Available at http://web.worldbank.org/WBSITE/EXTERNAL/TOPICS/EXTSOCIALDEVELOPMENT.

9 A sequential analysis of the welfare effects of mobile phones in Africa

Ten years ago, an article appeared which sought to reconcile the then-available concepts and numbers that bore on the status of mobile phones in Africa.[1] It distinguished, for example, between owning and nonowning users, between access and adoption and between participation and use. And judging by the relatively large number of citations that the article received,[2] there is some reason to think that it was successful, at least in part, in meeting its goals.

Since 2007, however, new concepts and data applications have made it possible to analyse these effects more closely and to move beyond the limits of the article published in that year. One of the main recognitions is the need for an analytical framework in which the welfare effects occur at different stages, with those at earlier points in time influencing those that occur later on. Much depends, for example, on the adoption stage, since the multiple forms that occur at this stage influence the ultimate welfare impact (such as when there are positive network effects that come into being when a certain degree of adoption has been reached).

I am also at pains to acknowledge that the traditional Western mode of measuring mobile subscribers does not survive close analytical or empirical scrutiny, for what it measures are the number of SIM card subscribers rather than the number of people with subscriptions (it is the latter, after all, that is important for welfare purposes). What is interesting, for such purposes, though, are the ways in which ownership of only SIM cards can be used to gain access to mobile phone use, as opposed to ownership. Such cards, for example, can be used in the mobile phones of family and friends (moreover, work on Ethiopia[3] reveals an active market for renting SIM cards).

More generally, what the chapter reveals is a wide range of activity in the way mobile technology and SIM cards are used in Africa, activity which often differs quite sharply from the conventional Western-based model. Perhaps the best example of this is the way in which the welfare effects of mobile phones as a means of communication vary according to the alternative ways of keeping in touch with friends and family. Because they generally enjoy few alternatives, poor developing countries generally get more out of mobile phones than richer countries. Considering only the fixed-line mode of communication as an alternative leads, in fact, to leapfrogging and the differential benefits derived there from (see below).

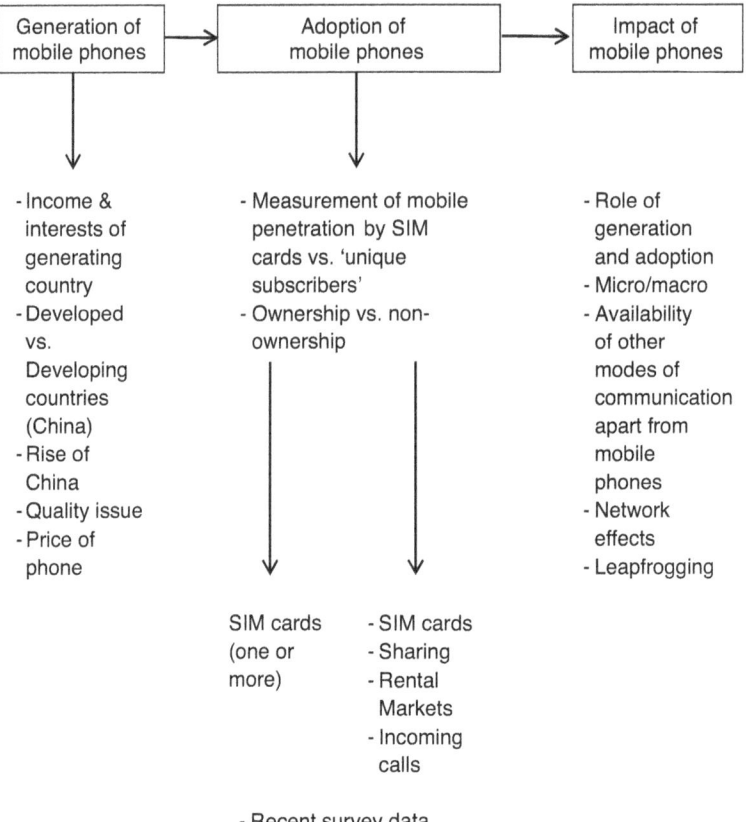

Figure 9.1 A sequential analytical framework

Source: James (2018)

Figure 9.1 presents the analytical framework as an outline of what is to follow in the rest of the chapter.

The first column deals mainly with the nature and price of mobile phones generated in China, as they now dominate many mobile markets in Africa. Under the second column fall the main differences in adoption between rich and poor countries, including ownership and non-ownership and the welfare effects of the latter in Africa (where many mobile devices and SIM cards are shared and rented rather than owned). The last column emphasises that these effects depend heavily on what alternative means of communication are available to users. This line of argument has a close affinity with the leapfrogging debate and the presence of fixed-line phones. It is not, as far as I am aware, an argument that has been advanced elsewhere in the literature (in the present form at least). Whereas some of the issues have already been raised, they have usually been discussed separately and in disparate places.

The generation of mobile phones

Whether or not mobile phone technology is suitable to African consumers depends to a large extent on where the technology is generated, for it is this locational issue that influences the price, quality and versatility of the product. In terms of the characteristics that they embody, the idea, as noted above, is that mobile phones imported from relatively rich countries will tend to be intensive in status, quality and consumer entertainment applications (such as social media). On the other hand, such phones tend to be inappropriate to those with much lower incomes, whose primary need is for basic communications, which until, say, 2015 were largely unavailable in Africa. According to Castells et al. (2007: 218), for example,

> In developed countries, mobile phones are defined by the term mobile and appreciated as a means to communicate on-the-go. However, the immediate benefit for people in developing contexts is that of connectivity, associated with having a means of communication whether mobile or not. Thus, considerations linked to mobility, style, and internet access, for example, are arguably secondary to basic connectivity at this stage . . . It is therefore important to identify the true source of developmental benefit. For people who already own a landline at home and/or at work, mobile phones bring the added benefit of mobility and convenience. To those for whom the mobile phone is the first form of personal communication to be owned, the major prize is to be connected at last; the phone is acquired not in order to be mobile, but in order to be connected, although mobility is an added bonus.
>
> (Castells et al., 2007: 218)

Data from a PEW study of selected African countries indicates that the majority of consumers derive their benefits mainly from the form of mobile telephony that yields the basic form of communications described in the quotation just provided. Thus, in Table 9.1, for example, the median result is that 65 per cent of the sample own basic mobiles i.e. those that are not smartphones. Only 15 per cent of the sample are able to afford these more expensive and sophisticated forms of communication, and a slightly higher percentage own no such technology whatsoever.

The powerful role of income in shaping these results is suggested by the extremes of Table 9.1, in conjunction with the country per capita incomes from Table 9.2. The two richest countries (South Africa and Nigeria) for example, enjoy the highest percentage rate of smartphone ownership, and these two countries also exhibit the lowest percentage of those without any mobile phones. At the opposite extreme lie Tanzania and Uganda, two of the three poorest countries, which exhibit both the lowest percentage amounts of smartphones and the highest percentage amounts of those without any mobile phones.

Table 9.1 Mobile phone distribution, selected African countries, 2015

Country	Ownership of smartphone, per cent	Ownership of mobiles other than smartphones, per cent	No mobile phone, per cent
South Africa	34	55	10
Nigeria	27	62	11
Senegal	15	69	17
Kenya	15	67	18
Ghana	14	69	17
Tanzania	8	65	27
Uganda	5	60	34
Median	15	65	17
USA★	64	25	11

★ Note: USA data from December 2014 PEW Research Center Surveys.

Source: PEW (2015)

Table 9.2 Per capita income, selected countries, 2015 ($) (descending order)

Country	Per capita income★
South Africa	13,195.50
Nigeria	6,003.90
Ghana	4,210.50
Kenya	3,088.80
Tanzania	2,672.50
Senegal	2,420.80
Uganda	1,851.00

★ Current international dollars.

Source: World Bank Indicators.

Note, too, that different types of mobile phones from China tend to be supplied to users in the relatively rich and poor countries contained in the table. The latter countries, for example, tend to rely on phones without an Internet connection and other sophisticated characteristics (such phones, indeed, can readily be purchased for less than $10 USD[4]). We should note that according to one point of view, that of a large Chinese multinational, its success in Africa has been due to the localisation of its products (China Daily, Africa, 2015). The latter is made possible by the establishment of special R&D centres for the region by the multinational. This may well have led to appropriate mobile phones that are both 'dustproof and resistant to high temperatures' (China Daily, Africa, 2015). Unfortunately, there is little evidence on how widespread this pattern of technological behaviour actually is. For countries such as South Africa and Nigeria, on the other hand, where incomes are substantially higher, low-cost smartphones from China tend to dominate (I am thinking here for example of brands such as Huawei & Lenovo). Not all developing countries have adequate supply-side capabilities to produce smartphones, but China is certainly among the group that does.[5]

To what extent the low cost of mobile phones from this country is due to the absence of certain characteristics or their relative low quality is difficult to say. But it is a question that needs to be answered in determining their welfare effects, all the more so because of the remarkably rapid rate of growth of Chinese mobile phones in the African market, during, say, the last ten years. For example, in 2003, the rate of subscriptions was only 60 per 1,000 inhabitants in Africa, whereas by 2013 the rate had increased to 742 for the same number of inhabitants, an increase of more than twelvefold.[6]

The adoption of mobile phones

Broadly, mobile phone adoption is the process of 'taking up' the innovations described in the previous section. But once one probes into this process somewhat more than superficially, it turns out to be much less simple than it seems, from a welfare point of view.

A major problem lies in the distinction between mobile phones subscribers defined in terms of actual human beings, as opposed to the number of SIM cards. The latter concept was relevant at a time 'when most mobile phone users had a single subscription and it was therefore statistically valid to assume that subscriptions equalled subscribers. However, as the price of handsets and services fell and prepaid services became popular and coverage ubiquitous, it became common . . . for users to have multiple SIM cards and mobile devices' (ITU, 2016: 158).

Indeed, it is legitimate to argue that 'The mobile–cellular-subscription indicator is thus becoming obsolete as it refers to registered SIM cards rather than people',[7] and it is after all with people that welfare effects are concerned. What is needed are measures of mobile phone ownership and use. These would need to be gathered by surveys which are a far more costly proposition than relying on SIM card numbers provided by telecommunication providers.

Not surprisingly, therefore, there is a lack of evidence on the relative size of ownership and use in different countries, but there is some. One small sample, of fourteen countries, for example, includes a group of African countries. Apart from a minority of countries where there is a small difference between ownership and use, in the majority of cases the disparity is relatively large (14 per cent points or more).[8] These majority cases (including some from Africa) suggest that many people access mobile–cellular services by sharing a device and/or a SIM card[9] (ITU, 2016: 165). The situation is particularly marked in India and Bangladesh, where around fifty per cent of users rely on someone else's SIM card or mobile device.

In the African context, it is well to note that prepaid SIM cards are being actively traded in some areas without the accompaniment of mobile handset ownership.

> In this respect a study on telecommunications in Ethiopia notes that there is an innovative market for selling SIM card rights and renting SIM cards . . .

For instance, there is a practice of owning more than one SIM card as a means of generating income. In addition the study describes an example of someone who owns three SIM cards, rents out two of them and uses the income to cover the costs of the third. In addition, the study describes that individuals who have just acquired their SIM card often keep the card for a while instead of using it, as they are still unable to buy a mobile phone or because they want to make optimum use of the airtime that is included in the prepaid subscription.

(Adam, 2005)

Note, in regard to nonowning users, that they often also make use of commercial rental markets for mobile phones. As argued below, these are wont to take different forms, ranging from kiosks in India and 'umbrella' phone sellers in African cities, to the strikingly successful village payphones in Bangladesh (where the mobile is used effectively as a fixed-line phone). With the notable exception of the last-mentioned scheme, these rental operations differ from sharing: first, in that they do not generally allow incoming calls, reducing in this way the welfare of those who can only make calls of an outgoing variety (sharing is often with family and friends who, one might suppose, would be willing to pass on information about incoming calls and to allow those who missed the calls to return them). As pointed out below, however, much depends on the degree of goodwill between the parties.

A second welfare difference between renting and sharing is also related to the degree of anonymity between the parties in the two cases. That is to say, whereas renting is explicitly commercial, sharing often takes place between family and friends who usually do not levy a charge for the exchange.

These and other examples boost my tentative conclusion that it is difficult to overstate the difference in impact of mobile phones in developing as against developed countries, and that such a difference is due to pronounced institutional disparities and differences in customs and culture between them. I shall continue to pursue this refrain in what follows and return to it in the conclusions to the chapter.

The impact of mobile phones

In assessing the overall welfare effects of mobile phones in Africa, a basic recognition is that such effects depend heavily on the two sequential phases of the project cycle described earlier, namely, generation and adoption. Indeed, within each of these phases, as already noted, important developments have occurred in the quite recent past that would significantly alter the way in which we think about the welfare effects of mobile phones in Africa (though the measurement issues related to these developments are still largely in a preliminary stadium).

I am thinking here in particular of the potential afforded by the increase in imports of mobile phones from China in many countries of the region, and also the distinction between subscriptions based on SIM cards vs those based on actual human involvement (otherwise known as 'unique' mobile subscribers). The import effect raises the prospect of higher welfare from the use of phones, while the distinction between SIM cards and unique mobile subscribers also has implications for well-being. For example, it seems clear to me that a given number of persons with a mobile subscription carries more weight than the same number of subscribers measured by SIM cards.

But the welfare effects of mobile technology are also influenced by factors that have been discussed in earlier chapters. These generally involve the presence or absence of alternative means of communication, including the availability of infrastructural alternatives (such as travel) to using the mobile phone as a means of communication. Generally speaking, the more plentiful, feasible and low-cost are the alternatives, the less will tend to be the positive impact of the mobile phone or SIM card (see below).

Consider this hypothesis, first, in relation to the growth effect that the new technology engenders. Table 9.3 shows the four studies (and their results) of the issue that I could find.[10] Three of them indicate that the impact of mobiles on growth is greater in poor than rich countries. Partly, this is a question of technology, and more specifically the notion of technological leapfrogging. This is the idea that poor countries can move directly to mobile phones without having first to adopt the previous technology, fixed-line phones, and the greater the lack of such competing technology, the larger tend to be the gains from the use

Table 9.3 The impact of mobile phones on economic growth, selected cases

Author(s)	Country coverage	Results
Sridhar and Sridhar (2004)	63 developing countries	'We find significant effects of . . . cell phone penetration on economic growth but lower than that found for OECD countries, dispelling the convergence hypothesis' (p. 25)
Lee et al. (2009)	44 sub-Saharan African countries	'We find that mobile cellular phone expansion is an important determinant of the rate of economic growth in Sub-Saharan Africa. Moreover, we find that the marginal impact of mobile telephone communication is even greater wherever land-line phones are rare' (p. 1)
Waverman et al. (2005)	92 high and low-income countries	'The growth dividend of increasing mobile phone penetration in developing countries is therefore substantial' (p. 11). Mobile phones have twice as large an impact on developing as developed countries.
Qiang (2009)	120 countries	'For every 10 percentage point increase in the penetration of mobile phones, there is an increase in economic growth of 0.81 percentage points in developing countries versus 0.60 percentage points in developed countries' (p. 8)

Source: James (2016)

of mobile phones (this is part of the reason, for example, why Waverman et al. (2005) find that mobiles have double the impact on growth in developing, as opposed to developed, countries).

According to PEW (2015), fixed-line penetration in the seven African countries surveyed was negligible. In fact, 'A median of only 2% across these nations say they have a working landline telephone in their house, with a median of 97% saying they do not have one. By contrast, 60% of Americans have a landline telephone in their household' (PEW, 2015: 7).

The argument advanced so far, however, has involved only the relative absence of fixed-line telephony in developing countries, and more specifically, the notion of technological leapfrogging. But the principle can also readily be generalised to include all other alternatives to mobile phones as a means of communications in developing countries. I am referring here, for example, to the availability of physical transport such as taxis, cars and public transport, public payphones and friends and family, with mobile or fixed-line phones. The idea is that the more scarce are these alternative modes of communication, the greater will tend to be the welfare effects of mobile phones (since they then contribute more to reducing the severe lack of communications that so typifies most poor African countries). To this intuitive line of reasoning, moreover, should be added the proposition that alternatives to the mobile phone are least in evidence among poor people in poor countries. This too requires little in the way of theoretical support, for a relative scarcity of goods and services itself almost constitutes a definition of underdevelopment.[11]

I will shortly present some micro-survey evidence on these issues for several African countries, but before that it is well to cite the main results of the Grameen Telecom initiative mentioned above. Recall, in this regard, what was already mentioned above: namely, that one welfare effect of the introduction of mobile phones in Bangladeshi villages was to increase the consumer surplus of the poor relative to the non-poor (see below). Table 9.4 suggests a reason why this would have been the case.

The entries show the responses by income group to the question 'How would you meet the purposes of the current calls [made with a rented village phone], had there been no VPP in your village?' Especially for the lowest income group,

Table 9.4 Alternatives to village payphones (VPPs) (per cent of respondents)

	Extremely poor	Moderately poor	Non-poor
Would not try	–	–	2.3
Telephone from other phone	26.5	36.5	43.0
Post office	5.9	7.1	6.8
Have to go/hire Person to go	67.6	56.4	47.3
Other	–	–	0.6
Total	100.0	100.0	100.0

Source: Bayes et al. (1999: 28)

the most common response to the question would have been the most onerous, namely, to either go to the location in question or hire someone else to do so. It is no wonder then that the poor should have gained most from the introduction of village phones. More generally,

> the poor usually do not have much in the way of alternatives to communicate with the outside world, neither relatives to help with a phone call, nor relatives to provide a ride to the destination. For the poor, the advent of VPPs opened up a lower-cost alternative for exchanging information.
>
> (Bayes et al., 1999: 29)

For the African case I rely heavily on the study of several countries in the region, by Samuel et al. (2005), because it is one of the most extensive survey contributions to the literature on the benefits of mobile phones. It is useful not just because of the breadth of the situations that are covered, but also because of the attention it pays to an extreme welfare effect of the positive kind – that is, one where without the technology, contacts with family and friends would be entirely prohibitive, as opposed to being merely very difficult. Moreover, the main effect of mobile phones was shown by the survey evidence to involve improved contact with friends and family.

Indeed,

> The main . . . impact identified by the surveys related to easier contact with family and friends. In both Tanzania and South Africa, many people move away from their home to find work, and mobile phones are now an important means of keeping in touch with families. In the survey sample, 91% of respondents in Tanzania called friends and relatives rather than travelling to see them. In South Africa, 77% of mobile users called rather than visited. Indeed, for many families surveyed the costs of travelling to see relatives would be prohibitive, especially in the poorest rural communities, and mobile phones represent the only option of maintaining contact.
>
> (Samuel et al., 2005: 49)

Another major dichotomy that bears on the form taken by adoption is ownership vs non-ownership of mobile phones and SIM cards. The distinction is salient because whereas ownership is the form most commonly adopted in rich countries, it is much less widespread in poorer nations (as shown in the chapters above). And in this latter group, the nature of 'taking up' mobile phones becomes a more complicated process with varying implications for welfare.

In the other extreme case, for example, where neither the phone nor the SIM card is owned, use of the former may still be possible in the form of sharing or rental markets. The former type of institution (sharing) is especially common among the relatively poor income groups in developing countries (see below) and is encouraged by the large average size of households there. Minges (2012: 116), for example, has usefully described the Senegalese case in this regard. In particular, his point is that 'Access is particularly high in countries with large households . . . This larger household size can dramatically extend access to mobile phones, considering that on average, nine persons are in each Senegalese household'.

According to the Measuring the Information Society Report of 2016, moreover,

> Studies conducted in developing countries in Africa and Asia show that numerous people may not have a subscription but still use mobile-cellular services by sharing someone else's subscription and/or phone. Other studies have shown that mobile phone sharing decreases as the percentage of phone owners increases . . . and that people at the bottom of the pyramid share their mobile phones mainly with family members; usually the male head of the household with the spouse. This suggests that mobile phones may be used as household devices in some contexts.
>
> (ITU, 2016)

The specific form of sharing that is mentioned in this citation carries important welfare implications. To the extent that it occurs, sharing between spouses is usually a relatively intimate phenomenon, one that would presumably make more likely the passing on of incoming calls meant for either of them. Much depends, though, on the degree of trust between the parties. Where this is relatively low, as in, say, 'bad' marriages, the degree of co-operation in dealing with incoming calls will also tend to be less and vice-versa.

Other findings by the same authors show that the costs of visiting family and friends in relation to the costs incurred in phoning them are dependent, as predicted in my earlier argument, on the existence and functioning of various modes of transport, such as roads and public services. For example, travel time and costs avoided by the mobile phone as a means of communication 'were slightly larger' for Tanzania, where roads are inferior and public transport less pervasive. Or again, some South African respondents cited the weak public phone network as a reason for relying on mobiles.

In general, then, the gains from mobile phones in Africa are the result of a substitution of this technology for physical travel under conditions of poor infrastructure, inadequate (and at times dangerous) forms of public transport and endemic ill health (an area where considerable gains can be observed under the heading of M-health). In many cases, indeed, mobiles are the only feasible forms of communication in poor, rural areas of the region. That is why there has been a general shift towards digital and away from physical means of communication in the region.

As Porter (2016) has rightly pointed out, however, there are social limits on how far this form of substitution can actually be taken. For example, at events such as funerals, close relatives may be expected to attend in spite of severe barriers to physical travel. In business as well, face-to-face contact between parties is sometimes necessary, to engender the trust that is so necessary to foster tight relationships of social capital. And in yet another example, it may be socially unacceptable to continually send money (via a scheme such as M-Pesa in Kenya) to friends and relatives without at least some amount of physical contact.

The general point is that physical presence imbues participants with benefits that are not (or are less) available from mobile phones alone. Thus: 'Sufficient physical travel is required to satisfy particular social obligations and to observe the rituals and sustained quality time often at particular moments and within specific kinds of ambient place, places appropriate for a certain affective quality' (Urry, 2012: 26). Or again,

> Given the fundamental differences between face-to-face interaction and electronically mediated exchange, this importance attached to face-to-face interaction is hardly surprising. – copresence in time and place allows a cycle of interruption, feedback, and repair that is virtually instantaneous. This has implications for negotiating identity, uncertainty, and ambiguity; reducing duplicity; and establishing and maintaining multidimensional, robust relationships: It might be essential for mobilizing collective action in situations of uncertainty and risk. The phone, by contrast, might filter out not only social context clues such as location but the full range of psychoemotional reactions, such as discomfort or attraction.
>
> (Porter, 2016: 434)

The final entry on impact in the last column of Table 9.1 above refers to so-called network effects. These refer to the gains experienced by an individual subscriber of mobile phones, as the number of other subscribers increases. Here, as in the case of leapfrogging discussed earlier, and for basically the same reason, there is an important difference in the gains experienced by rich and poor countries. In particular, according to Castells et al. (2007: 216) 'Mobile phones provide significantly higher network effects in developing than in developed countries where fixed lines *have already performed this function*'. Or, to put the point another way, it is that 'while in developing countries the benefits of mobile are two-fold-the increase in the network effect of telecoms *plus* the advantage of mobility-in developed economies the first effect is much more muted' (Waverman et al., 2005: 17).

Apart thus from the wide spread of the village phones, there are two other welfare effects of the scheme that deserve particular emphasis. One of them has to do with the role of incoming calls and the willingness of phone operators to alert the villager when he or she receives, but misses, an incoming call (family members of the village 'phone ladies', for example, are sometimes used for this purpose). The value to the intended recipient of the call, however, differs quite sharply from one case to another, depending on its content. This may range, for example, from a routine social call on the one hand, to an important message about work on the other. In the latter event, there is likely to be a notable welfare difference between the Grameen project and most commercial sharing arrangements, which do not allow incoming calls.

The second notable aspect of the project from our point of view is its distributional outcome, as measured by the consumer surplus (CS), of different

Table 9.5 Consumer surplus by income group, selected Bangladesh villages (in Taka)

Economic status	Total costs of alternative methods (1)	Total costs of village payphones (2)	Consumer surplus (3) = (1) – (2)
All poor	95.21	17.35	77.86
Extremely poor	81.38	20.08	61.30
Moderately poor	106.71	15.07	91.64
Non-poor	67.51	16.73	50.78

Source: Bayes et al. (1999: 30)

income groups (where the CS is measured by the difference between what a user is prepared to pay for mobile services and what he or she actually pays). As shown above, the consumer surplus for the poor is considerably higher than the non-poor, as shown in Table 9.5.

Conclusions

There are certainly studies on the impact of mobile phones on the welfare of those living in the African region. Generally speaking, however, they tend to be written in isolation from one another. A major purpose of this chapter, accordingly, has been to redress the fragmented nature of the relevant literature.

The synthesis thus attempted is based on a sequential analytical framework, within which impacts at one stage influence what occurs in later stages (whether, for example, mobile phones are made in China or America determines the prices and characteristics of the product and hence the welfare impact on different groups).

Within the chosen framework, several themes emerged that bear reiteration. One of them concerns the need in Africa to transcend the basic Western model, where both the mobile instrument and the SIM card are owned by the individual user. Where, for example, there is ownership of only the SIM card, the poor user may be able to use this in someone else's mobile or to buy more than one card and rent them out. Then, there are various forms of commercial renting out of mobiles themselves, such as 'umbrella' shops on many African city street corners.

A very different (though related) form of institution involves non-commercial sharing between individual friends and family (often between spouses in large households). Such a cooperative form of behaviour, while by no means inevitable, seems appropriate to those with low incomes and indeed seems to fit in well with what some describe as a 'culture of sharing' in Africa (though this view is not universally accepted).

Another pervasive theme of the chapter – one of its most novel – is that the welfare effects of mobile phones in Africa depend on the ubiquity of alternative forms of communication. The hypothesis is (or was) that the less available are the alternatives to mobile phones, the greater will be the gains from the phones themselves. This hypothesis was widely confirmed at the level of both households and countries. When it is applied specifically to fixed-line phones, the argument takes the form of leapfrogging: that the lack of this type of phone enables poor countries to move directly to the new technology, deriving higher growth rates than in richer countries (see also the higher consumer surplus of individuals most lacking in alternatives to the mobile phone such as public transport services, public phones and so on). Thus, according to Table 9.4 above, the form taken by the adoption of mobile phones tends to have a profound effect on the impact that they contribute.

Notes

1 James and Versteeg (2007).
2 According to the SSCI, the number of citations in July 2017 was equal to 160.
3 See discussion below.
4 My source here is www.alibaba.com, July 2017, a famous Chinese-based website.
5 One can gauge this, for example, by studying the percentage of high-tech exports from China over the years. See yearbooks of International Trade Statistics, over, say, the past fifteen years.
6 Eurostat (2015). The World Bank (data) has a slightly different result because it excludes high-income countries from sub-Saharan Africa.
7 SIM card data are routinely recorded by phone providers and are thus readily available (for measuring subscriptions per card by country).
8 See ITU (2016).
9 I was helped here by what the World Bank had discovered. See Qiang (2009).
10 Scarcity economics is sometimes known as neoclassical economics, which certainly captures part of the overall problem of development. But it also ignores many other relevant issues, not the least of which is distribution of income and wealth.
11 Grameen is also unusual with respect to the empowerment of its female mobile phone operators, whose status in the village was said to have increased noticeably. Many countries of Africa, by contrast, are known to exhibit gender bias (see ITU, 2016) in their access to and use of mobile phones.

References

Adam, L. (2005). Ethiopia, in A. Gillwald (ed.) *Towards an African e-Index: ICT Access and Usage*, Johannesburg: Witwatersrand University School of Public and Development Management, the LINK Centre.

Bayes, A., Von Braun, I. and Akhter, R. (1999). Village pay phones and poverty reduction: Insights from a Grameen Bank Initiative in Bangladesh, ZEF discussion paper on Development Policy, number 8, Bonn.

Castells, M., Fernandez-Ardevol, M., Qui, T. and Sey, A. (2007). *Mobile Communication and Society: A Global Perspective*, Cambridge, MA: MIT Press.

China Daily, Africa (2015). *Chinese Phone Makers Got Smart in Africa*. Available at www.chinadaily.com.cn/weekly/2015.../content_19989393.

Eurostat (2015). Africa-European Union, *Key Statistical Indicators*.

ITU (International Telecommunication Union) (2016). *Measuring the Information Society*, ITU: Geneva.

James, J. (2016). *The Impact of Mobile Phones on Poverty and Inequality in Developing Countries*, Heidelberg: Springer.

James, J. and Versteeg, M. (2007). Mobile phones in Africa: How much do we really know?, *Social Indicators Research*, 84:117–126. Doi:10.1007/S11205-006-9097-x.

Lee, S., Levendis, J. and Gutierrez, L. (2009). *Telecommunications and Economic Growth: An Empirical Analysis of Sub-Saharan Africa*. Available at http://papers.ssrn.com/5013/papers.cfm?abstract_id=1567703.

Minges, M. (2012). Overview, in *World Bank*, Washington, DC: Maximizing Mobile.

PEW (2015). Cell Phones in Africa: Communication Lifeline, April. http://www.pewglobal.org/2015/04/15/cell-phones-inAfrica-

Porter, G. (2016). Mobilities in rural Africa: New connections, new challenges, *Annals of the American Association of Geographers*, 106,2:434–441. doi.org/10.1080/00045608.205.110 0056.

Qiang, C. (2009). *Telecommunications and Economic Growth*, Washington, DC: The World Bank.

Samuel, I., Shah, N. and Hadingham, W. (2005). Mobile communications in South Africa, Tanzania and Egypt: Results from community and business surveys, *Vodafone Policy Paper Series*, 2:44–52. Available at www.vodafone.com/content/dam/vodafone/about/public_policy/policy.papers/public_policy_series_2.pdf. Accessed September 10, 2013.

Sridhar, K. and Sridhar, V. (2004). Telecommunications Infrastructure and Economic Growth: Evidence from Developing Countries, *Indian National Institute of Public Finance and Poverty*. Available at http://econpapers.repec.org/paper/npfwpaper/ox_2f14.htm.

Urry, J. (2012). Social networks, mobile lives and social inequalities, *Journal of Transport Geography*, 21:24–30. doi.org/10.1016/j.jtrangeo.2011.10.003.

Waverman, L., Meschi, M. and Fuss, M. (2005). The impact of telecoms on economic growth in developing countries, *Vodafone Policy Paper Series*, 2:10–24. Available at www.vodafone.com/content/dam/vodafone/about/public.policy/policy.papers/ppublic_policy_series.2.pdf. Accessed July 6, 2014.

World Bank, data (2013). Washington, DC.

World Bank Indicators (2015). Washington, DC.

Index

Note: Page numbers in italic indicate figures and in bold indicate tables on the corresponding pages.